SpringerBriefs in Applied Sciences and Technology

Computational Intelligence

Series Editor

Janusz Kacprzyk, Systems Research Institute, Polish Academy of Sciences,
Warsaw, Poland

W0234394

SpringerBriefs in Computational Intelligence are a series of slim high-quality publications encompassing the entire spectrum of Computational Intelligence. Featuring compact volumes of 50 to 125 pages (approximately 20,000-45,000 words), Briefs are shorter than a conventional book but longer than a journal article. Thus Briefs serve as timely, concise tools for students, researchers, and professionals.

Jyotismita Chaki

The Art of Deep Learning
Image Augmentation:
The Seeds of Success

 Springer

Jyotismita Chaki
School of Computer Science
and Engineering
Vellore Institute of Technology
Vellore, Tamil Nadu, India

ISSN 2191-530X   ISSN 2191-5318 (electronic)
SpringerBriefs in Applied Sciences and Technology
ISSN 2625-3704   ISSN 2625-3712 (electronic)
SpringerBriefs in Computational Intelligence
ISBN 978-981-96-5080-4   ISBN 978-981-96-5081-1 (eBook)
https://doi.org/10.1007/978-981-96-5081-1

This Springer imprint is published by the registered company Springer Nature Singapore Pte Ltd.
The registered company address is: 152 Beach Road, #21-01/04 Gateway East, Singapore 189721, Singapore

If disposing of this product, please recycle the paper.

Preface

In the realm of deep learning, the adage "garbage in, garbage out" rings particularly true. The performance of computer vision models hinges heavily on the quality and quantity of the training data. Limited data poses significant challenges, leading to overfitting, poor generalization, and difficulties in learning complex concepts. This book delves into the art of image augmentation, a powerful technique that addresses these challenges by artificially expanding training datasets.

I begin by exploring traditional augmentation methods, such as geometric transformations and color space manipulations, while acknowledging their limitations. I then embark on a journey into the exciting world of deep learning-based augmentation, showcasing how techniques like GANs and autoencoders can generate highly realistic and diverse synthetic images.

The book delves into the practical aspects of image augmentation, covering key applications, evaluation strategies, and optimization techniques. I explore how augmentation enhances various computer vision tasks, from object detection and image segmentation to image-to-image translation. I discuss effective methods for evaluating the impact of augmentation on model performance, including both quantitative metrics and qualitative assessments. Furthermore, I delve into strategies for optimizing augmentation pipelines, such as hyperparameter tuning and the development of adaptive augmentation policies.

Finally, I explore cutting-edge advancements in the field, including AutoAugment, interpretable augmentation, and attention-based methods. I also introduce the concept of Human-in-the-Loop Augmentation, where human expertise is integrated into the augmentation process, leading to more robust and trustworthy models.

This book aims to provide a comprehensive overview of deep learning image augmentation, empowering researchers and practitioners with the knowledge and tools to effectively use this powerful technique. By mastering the art of image augmentation, we can unlock the full potential of deep learning models and drive significant advancements in computer vision and beyond.

Vellore, India   Dr. Jyotismita Chaki

Contents

About the Author

Dr. Jyotismita Chaki is an associate professor at the School of Computer Science and Engineering, Vellore Institute of Technology, India. She holds a Ph.D. in Engineering from Jadavpur University, Kolkata, and her research interests encompass Computer Vision, Image Processing, Pattern Recognition, Medical Imaging, Artificial Intelligence, and Machine Learning. Dr. Chaki is an author and editor, with a substantial body of work including five authored books published by renowned presses like Springer and CRC Press, and six edited books published by CRC Press and Elsevier. She has also published many research articles in high-impact, SCIE-indexed journals, the majority of which are ranked in the top quartiles (Q1 and Q2). In recognition of her contributions, Dr. Chaki was named the world's top 2% scientist by Stanford University and Elsevier in 2024. She is also a Senior Member of the IEEE. Dr. Chaki's editorial contributions are extensive, currently serving as editor for 9 journals, including *Engineering Applications of Artificial Intelligence* (Elsevier), *Scientific Reports* (Nature Portfolio), *Discover Applied Sciences* (Springer Nature), *PLOS ONE*, *PeerJ Computer Science, Computer and Electrical Engineering* (Elsevier), *Array* (Elsevier), *Machine Learning with Applications* (Elsevier), and *BMC Artificial Intelligence.*

Chapter 1
Introduction to Deep Learning-Based Image Augmentation

Deep learning models in computer vision often rely on large datasets for optimal performance. However, acquiring and annotating extensive image datasets can be time-consuming and expensive. To overcome this limitation, data augmentation techniques have emerged as crucial tools. By artificially expanding the existing dataset through various transformations, image augmentation significantly enhances the training process, improves model robustness, and ultimately leads to better performance. This chapter will delve into the intricacies of deep learning-based image augmentation, exploring advanced techniques that go beyond traditional methods and unlock new possibilities for improving model accuracy and generalization.

1.1 Challenges of Limited Images in Deep Learning

Limited image datasets pose several significant challenges for deep learning models in computer vision [1].

When training data is limited, deep learning models face the significant challenge of overfitting. Overfitting occurs when the model memorizes the specific details and noise present in the training data instead of learning the underlying patterns and variations that generalize to unseen data. With insufficient data, the model may struggle to identify and learn the essential features and relationships within the images. Consequently, the model may accomplish exceptionally fine on the training set but poorly on unseen, new images, as it cannot generalize its knowledge to novel situations. This overfitting behavior significantly hinders the model's ability to effectively perform in real-world scenarios where the data distribution may differ from that of the training set.

Limited image datasets restrict the model's exposure to the diverse variations that exist in real-world scenarios. Real-world images exhibit significant variability. Variations in ambient lighting, shadows, and glare can drastically alter the appearance

J. Chaki, *The Art of Deep Learning Image Augmentation: The Seeds of Success*, SpringerBriefs in Computational Intelligence, https://doi.org/10.1007/978-981-96-5081-1_1

of objects within an image. Objects can be captured from various angles, leading to significant changes in their appearance. Objects may be partially or fully occluded by other objects, obscuring parts of their visual information. Real-world images often contain complex backgrounds that can distract the model and make object recognition more challenging. When trained on a limited dataset, the model may not encounter a sufficiently broad range of these variations. As a result, it may struggle to generalize its knowledge to unseen images that deviate from the limited set of variations present in the training data. This lack of exposure to real-world variability can cause bad performance on unseen images, hindering the model's robustness and real-world applicability.

Deep learning models excel at identifying complex patterns and relationships within large datasets. However, when presented with limited training data, their ability to grasp intricate concepts is significantly hindered. With insufficient examples, the model may struggle to discern subtle variations, subtle nuances in object appearance, or complex relationships between different visual elements. For instance, accurately classifying fine-grained categories like bird species or recognizing subtle facial expressions often requires the model to learn intricate details and subtle variations. Limited data can hinder the model's ability to extract these complex features, leading to poor performance and an inability to distinguish between closely related classes or recognize subtle nuances in the input images. This limitation emphasizes the crucial role of large and diverse datasets in enabling deep learning models to effectively learn and represent complex visual concepts.

While limited datasets might initially seem computationally cheaper to train on, this is often not the case. Models trained on small datasets often require significantly more training epochs (iterations over the entire dataset) to achieve acceptable performance. This prolonged training process translates to increased computational costs. With fewer training examples, the model takes longer to converge to an optimal solution. Each epoch requires the model to process the entire dataset, and with limited data, the model needs to iterate through the dataset multiple times to learn effectively. This significantly increases the overall training time, requiring more computational resources and potentially delaying the model development process. Prolonged training time directly translates to increased computational costs. Training deep learning models requires significant computational power, often utilizing powerful GPUs or TPUs. Extended training times on limited datasets can lead to significantly higher energy consumption and computational costs compared to training on larger datasets that converge more quickly. Therefore, while limited data might seem to reduce the initial data collection and preparation efforts, the increased training time and computational costs associated with achieving acceptable performance can offset these initial savings.

Limited image datasets often suffer from inherent biases, where certain categories or sub-groups are overrepresented while others are underrepresented. For example, a facial recognition dataset might contain a disproportionate number of fair individual images, while images of people with darker skin tones are underrepresented. This bias in the training data can meaningfully influence the model's performance and lead to inaccurate or unfair classifications. When trained on such biased data, the model tends to learn and rely on features that are overrepresented in the training set. As a result, the model may exhibit significant performance disparities across different sub-groups. In the case of facial recognition, a model trained on a biased dataset might exhibit higher accuracy for individuals with lighter skin tones while demonstrating significantly lower accuracy for individuals with darker skin tones. This bias can have serious implications, leading to unfair or discriminatory outcomes in real-world applications. Furthermore, limited datasets may not sufficiently denote the assortment of real-world scenarios. For example, a dataset of self-driving car images collected primarily in urban environments might not adequately represent rural driving conditions. This can cause poor performance and safety issues when the model is installed in real-world scenarios that differ significantly from the training environment. Addressing these biases and ensuring fair and equitable performance necessitates careful consideration of data collection, curation, and augmentation strategies.

Deep learning models, mostly those trained on limited data, can often operate as "black boxes." This means that it is difficult to understand the internal workings of the model and how it arrives at its predictions. The complex, multi-layered architecture of these models, with numerous interconnected nodes and parameters, makes it challenging to decipher the reasoning behind their decisions. This absence of explainability poses several significant challenges. If we cannot understand how a model arrives at a particular decision, it becomes difficult to trust its predictions, especially in critical applications like healthcare or autonomous vehicles. Without understanding the model's reasoning, it is challenging to identify and address potential biases that may be present in the model's decision-making process. When a model makes incorrect predictions, it is difficult to pinpoint the source of the error. Without understanding the model's internal workings, it becomes challenging to debug and improve its performance. This lack of transparency hinders the widespread adoption and deployment of deep learning models in many critical applications. To overcome these challenges, there is a growing emphasis on developing more explainable and interpretable AI systems. This involves developing techniques that can provide insights into the model's decision-making process, allowing us to understand how the model arrives at its predictions and identify potential biases or weaknesses.

Limited datasets can exacerbate ethical issues like bias and fairness. When training data primarily reflects a specific demographic, the model may inadvertently learn and perpetuate existing societal biases. For example, a sentiment analysis model trained on a limited dataset of social media posts might predominantly reflect the language and sentiment expressions of a particular cultural or demographic group. This can lead to biased interpretations of sentiment, potentially misinterpreting or misclassifying the sentiment expressed by individuals from other groups. Furthermore, limited data can fail to capture the nuances and complexities of language used by different groups, including slang, idioms, and cultural references. This can result in the model misinterpreting or misclassifying sentiment expressed in these nuanced ways, leading to unfair or discriminatory outcomes. For example, a sentiment analysis model trained on a limited dataset might fail to accurately interpret sarcasm or irony, which can vary significantly across different cultural and linguistic groups. This can lead to inaccurate and potentially harmful interpretations of user sentiments, particularly for marginalized or underrepresented groups. Addressing these ethical concerns requires careful consideration of data collection, curation, and model development practices. It is crucial to ensure that the training data is diverse, representative, and free from biases. Additionally, ongoing monitoring[1] and evaluation of model performance across different demographic groups are essential to identify and mitigate any potential biases.

Limited image quality presents a significant challenge for deep learning models. High-quality images with sharp details and clear features are crucial for optimal model performance. However, real-world datasets often contain images with low resolution, blurriness, noise, or other artifacts. These imperfections can hinder the model's ability to accurately extract meaningful features and can lead to degraded performance. For example, a blurry image of a car might make it difficult for the model to accurately identify its make and model, leading to misclassifications. Furthermore, accurate annotations are crucial for training effective deep learning models. Annotations provide the ground truth information that the model learns to associate with the visual features in the image. However, limited datasets often come with limited or inaccurate annotations. This can be due to various factors, such as human error during manual annotation, inconsistencies in annotation guidelines, or the complexity of the annotation task itself. Inaccurate annotations can mislead the model, leading it to learn incorrect associations and ultimately hindering its performance. For instance, if the bounding boxes around objects in an image are inaccurate, the model will learn to associate the predicted object with an incorrect region, leading to incorrect detections. Therefore, ensuring high-quality images and accurate annotations is crucial for training effective and reliable deep learning models in computer vision.

Deep learning models trained on limited data can be particularly vulnerable to adversarial attacks. These attacks involve carefully crafted inputs that can subtly manipulate the model's predictions, causing it to make incorrect or unintended decisions. When trained on a limited dataset, the model may not have learned a robust and diverse representation of the target classes. This lack of robustness can make it easier for attackers to exploit subtle vulnerabilities in the model's decision-making procedure. Adversarial attacks can take various forms, such as adding invisible noise to an image, slightly modifying the input data, or even introducing subtle distortions that are imperceptible to the human eye. These seemingly minor perturbations can significantly mislead the model, causing it to misclassify objects, make incorrect predictions, or even reveal sensitive information. For example, in image classification, an attacker might introduce subtle, imperceptible noise to an image of a stop sign, causing the model to misclassify it as a speed limit sign. This could have serious consequences in self-driving car applications, potentially leading to accidents. The vulnerability to adversarial attacks is a significant concern, particularly in safety–critical applications. Therefore, it is crucial to develop robust defenses against such attacks, including techniques like adversarial training, data augmentation, and model regularization.

These challenges underscore the critical importance of large and diverse datasets for training effective deep learning models in computer vision. Techniques like data augmentation play a vital role in extenuating these challenges by effectively expanding the size and diversity of the training data.

1.2 Image Augmentation

Image augmentation is a crucial technique in computer vision that involves artificially expanding the size and diversity of a training dataset. By applying various transformations to existing images, such as rotations, flips, crops, color adjustments, and noise injections, we can create new, slightly modified versions of the original images. This synthetically expanded dataset exposes the model to a wider range of variations in the input data, improving its ability to generalize and perform robustly in real-world scenarios. For example, by rotating images during training, a model can learn to recognize objects regardless of their orientation, making it more robust to variations in object positioning. Similarly, adjusting brightness and contrast can help the model generalize to images with varying lighting conditions. In essence, image augmentation helps to make the model more adaptable and less susceptible to overfitting the precise features of the original training data. Here's a detail of its purpose and benefits [2].

Increase dataset size: Large datasets are decisive for training effective deep learning models, especially in computer vision. However, acquiring and annotating massive amounts of real-world data can be laborious, expensive, and sometimes even infeasible. Image augmentation addresses this challenge by artificially expanding the size of the training dataset. By applying various transformations to existing images, such as rotations, flips, crops, and color adjustments, we can generate numerous variations of each original image. This effectively increases the number of training samples available to the model without requiring any additional data collection. This expanded dataset exposes the model to a wider range of variations in the input data, improving its ability to generalize and perform robustly on unseen images. For example, if the model is trained only on images of cats facing the camera, augmentations like rotations can introduce images of cats facing different directions, improving the model's ability to recognize cats regardless of their orientation.

Improve model generalization: Image augmentation significantly improves model generalization by exposing the model to a wider range of variations in the input data. Real-world images exhibit important diversity in terms of lighting conditions, object orientations, scales, and viewpoints. For example, an object might appear differently under different lighting conditions (e.g., shadows, glare), be captured from various angles, or appear at different scales within the image. By applying augmentations like random rotations, cropping, scaling, and brightness adjustments, we simulate these real-world variations in the training data. This forces the model to learn more generalizable and robust features that are invariant to these transformations. As a result, the model becomes less sensitive to specific viewpoints or lighting conditions and can better recognize objects in unseen images with varying characteristics. In essence, augmentation helps the model learn to extract essential features that are robust to common variations, leading to improved performance on real-world data.

Reduces data collection and labeling costs: Collecting and labeling large datasets can be a significant bottleneck in deep learning projects. It often requires substantial time, effort, and financial resources. Image augmentation provides a cost-effective solution to this challenge. Generating numerous variations of existing images effectively amplifies the training dataset size without the need for additional data collection and labeling efforts. This significantly decreases the time and resources essential to gather and annotate new data, making it a more efficient and economical approach to training deep learning models. For example, instead of collecting thousands of images of cars from various angles, we can augment a smaller dataset of car images by applying rotations, translations, and scaling, effectively increasing the dataset size without incurring the costs associated with acquiring and labeling new images.

Helps address class imbalance: Image augmentation can effectively address class imbalance in datasets. When certain classes have meaningfully fewer images than others, the model is inclined to be biased toward the majority classes, leading to poor performance in the underrepresented classes. By applying augmentation techniques more aggressively to images belonging to the underrepresented classes, we can artificially increase their representation in the training set. For instance, if a dataset comprises many images of dogs but only a few images of cats, we can apply more extensive augmentations (e.g., more flips, rotations, and zooms) to the cat images. This effectively increases the number of "cat" images seen by the model, helping it learn to recognize cats more effectively and improve its overall performance on the classification task. This approach helps to mitigate the bias toward the majority classes and ensures that the model can accurately classify even rare or uncommon objects within the dataset.

Can be tailored to specific tasks: Image augmentation techniques can be carefully tailored to the specific computer vision task. For example, in object detection, where the precise location and size of objects are crucial, techniques like random cropping and occlusions can significantly improve model robustness. By randomly cropping different regions of the image, the model learns to detect objects even when they appear in different parts of the image or are partially occluded. This helps the model become less reliant on the object's position within the image and improves its ability to localize objects accurately. Simulating partial occlusions by randomly blocking parts of the image forces the model to learn robust features that are not easily disrupted by partial obscurations. This is particularly beneficial for real-world scenarios where objects might be partially hidden by other objects or by shadows. For tasks like pose estimation or action recognition, augmentations like rotations and shearing can help the model learn to identify objects or actions irrespective of their perspective or orientation. By carefully selecting and applying augmentation techniques that are relevant to the specific task, we can significantly improve the model's performance and its ability to generalize to real-world scenarios.

Improves data security: For datasets containing sensitive information, such as medical images with patient identities or facial images with personally identifiable information, data privacy is a serious concern. Augmentation can play a vital role in enhancing data security. By applying privacy-preserving augmentations, we can generate synthetic variations of the original data that maintain the essential characteristics for training while significantly dropping the risk of identifying individuals. For instance, in medical imaging, augmentations like slight blurring, noise injection, or small random rotations can effectively obfuscate fine-grained details while preserving the overall anatomical structures. This can make it significantly harder to re-identify individuals from the augmented images. Similarly, in facial recognition, techniques like adding noise, changing lighting conditions, or applying subtle distortions to facial features can make it difficult to recognize specific individuals while still preserving the essential features for facial recognition. By utilizing these privacy-preserving augmentation techniques, researchers and developers can use the benefits of data augmentation while minimizing the risks associated with handling sensitive data. This not only enhances data security but also promotes the ethical and responsible use of artificial intelligence in sensitive domains.

Reduces overfitting: Overfitting happens when a model memorizes the training data too well, performing exceptionally well on the training set but poorly on unseen data. This occurs because the model has learned to exploit specific patterns or noise within the training data that are not representative of the real-world distribution. Image augmentation helps to mitigate overfitting by introducing variations that force the model to learn more generalizable features. By applying transformations such as rotations, flips, crops, and color adjustments, we present the model with slightly modified versions of the original images. This prevents the model from over-relying on specific details or artifacts in the training data and encourages it to learn more robust and generalizable features that are invariant to these variations. For instance, if the model overfits the exact pixel values of a specific image, applying random noise or slight blurring can force the model to learn more robust feature representations that are less sensitive to minor variations in pixel intensities. This ultimately leads to better generalization performance on unseen data.

Improves model robustness: In real-world applications, images rarely appear under ideal conditions. Factors like varying lighting conditions (e.g., shadows, glare), rotations due to camera angles, slight occlusions from other objects, and changes in scale are common. By introducing these variations through augmentation, we expose the model to a more realistic and challenging training environment. For instance, by applying random brightness adjustments, the model learns to recognize objects even under varying lighting conditions. Similarly, by simulating rotations and slight occlusions, the model becomes more robust to changes in object orientation and partial visibility. This enhanced robustness allows the model to generalize better to unseen images that may exhibit similar variations, leading to improved performance and reliability in real-world scenarios.

Promotes creativity in data generation: Beyond simple transformations like rotations and flips, image augmentation can be used to generate entirely new images based on existing ones, pushing the boundaries of data generation. This "creative" aspect of augmentation opens up exciting possibilities. For instance, techniques like

Generative Adversarial Networks (GANs) can be employed to generate realistic variations of input images, such as altering hairstyles, changing clothing styles, or even modifying facial expressions. This can be particularly valuable for tasks like image inpainting, where the goal is to fill in damaged or missing portions of an image. By training a GAN on an image dataset, it can learn to generate plausible completions for missing parts, effectively creating new, synthetic images that are consistent with the original data distribution. This creative approach to data generation not only expands the training dataset but also allows for the exploration of novel image variations and the development of more robust and versatile image processing models.

Keypoint Detection: Keypoint detection tasks, such as human pose estimation, aim to accurately locate specific points of interest within an image, like joints in a human body. In these scenarios, subtle deformations of the object can significantly impact the appearance of the key points. To address this, a powerful technique is to apply random elastic deformations during image augmentation. Elastic deformations introduce localized, non-rigid transformations to the image. This involves randomly displacing small regions of the image in a controlled manner, simulating slight deformations or distortions. These deformations can mimic real-world variations such as slight bending, stretching, or compression of the object. By training the model on images with these elastic deformations, the model learns to be more robust to such variations. For example, in human pose estimation, applying elastic deformations to images of people can simulate slight bending of limbs, changes in body posture, or variations in clothing that might slightly distort the appearance of the body. This forces the model to learn the underlying relationships between key points even when the object undergoes subtle deformations, improving its accuracy and robustness in real-world scenarios where perfect rigidity is rarely observed.

1.3 Traditional Image Augmentation

Traditional image augmentation techniques are a set of methods for artificially expanding a dataset of images by creating variations of existing ones. These variations help improve the performance of deep learning models in computer vision tasks by addressing the limitations of limited data. Here are some of the commonly used techniques.

1.3.1 Geometric Transformations: Geometric transformations are a fundamental category of image augmentation techniques used to manipulate the spatial arrangement of pixels in an image. These manipulations introduce variations in object size, position, and orientation, mimicking real-world scenarios and enhancing the robustness of deep learning models in computer vision tasks. Geometric transformations are computationally efficient and relatively simple to implement. However, it's crucial to choose the right transformations and control the degree of manipulation to avoid introducing

unrealistic distortions or losing important information from the image [3].
Here's a detail of some key geometric transformations.

1.3.1.1 Rotation: Rotation is a fundamental image augmentation technique that
involves virtually rotating the entire image around its center point by a
specific angle. This manipulation introduces variations in object orientation and viewpoint, enhancing the robustness of deep learning models in
computer vision tasks. Here's how it benefits model training: (a) Viewpoint
Invariance: Real-world objects can be viewed from different angles. Rotation introduces variations that help the model learn to recognize the object
regardless of its orientation. This is crucial for tasks like object classification
(identifying a dog) or object detection (finding a car) where the object's pose
can vary significantly. (b) Addressing Limited Data: For tasks with limited
datasets, rotation can artificially expand the data by creating new images
with different rotations. This helps the model learn a more comprehensive representation of the object and reduces the risk of overfitting on the
limited training data. This helps the model recognize the object regardless of
its orientation. Rotation is a relatively simple technique to implement and
computationally efficient. The angle of rotation can be chosen randomly
within a predefined range to create diverse variations. However, excessive
rotation can lead to objects being cut off or appearing unnatural. It's important to find a balance that captures real-world viewpoint variations without
introducing unrealistic distortions. Figure 1.1 represents some examples of
rotated image samples.

1.3.1.2 Scaling: Scaling is a geometric image augmentation technique that modifies
the size of an image. This manipulation introduces variations in object scale
and can be crucial for training deep learning models in computer vision
tasks. Here are some of its impacts: (a) Simulating object distances: In
real-world scenarios, objects can be present at varying distances from the
camera. Scaling an image up or down creates new images where the object
appears larger (closer) or smaller (farther away). This helps the model learn
to recognize objects irrespective of their size in the image frame. (b) Data
Augmentation for specific tasks: For object detection tasks, scaling can be
particularly beneficial. By introducing scaled versions of objects, the model

Fig. 1.1 Oriented augmented image samples created from one image

Fig. 1.2 Scaled augmented image samples created from one image

learns to detect objects of different sizes with similar accuracy. This is crucial for real-world applications where object sizes can vary significantly. This helps the model recognize the object even if it appears at different distances in the image. There are various scaling strategies, including scaling the entire image by a fixed factor or using a random scaling factor within a defined range. Though, it's crucial to avoid excessive scaling that might cause a loss of information or create unrealistic object sizes. Figure 1.2 represents different scaled images.

1.3.1.3 Flipping (Horizontal/Vertical): Flipping, also known as mirroring, is a simple yet effective image augmentation technique in computer vision. It involves creating new images by reversing the pixels along a specific axis. Images are flipped horizontally or vertically. The horizontal flipping technique mirrors the image along the vertical axis, essentially creating a left–right mirrored version. It's particularly useful for tasks where objects can appear facing either direction (like pedestrians or animals). By introducing horizontally flipped versions, the model learns the concept of the object independent of its orientation. In vertical flipping, the image is mirrored along the horizontal axis, resulting in a top–bottom mirrored version. This augmentation can be beneficial for tasks where the object's natural orientation is important but slight variations might exist (like handwritten digits or airplanes). It helps the model become more robust to minor variations in vertical positioning. This accounts for situations where the object might be facing the other direction in real-world scenarios. Flipping is computationally cheap and easy to implement. However, it's important to consider the task at hand. For tasks where the object inherently has a specific orientation (like a car with a windshield), flipping might not be a suitable augmentation strategy. Figure 1.3 represents different flipped image samples.

1.3.1.4 Cropping (Random/Center/Specific): Cropping is a fundamental image augmentation technique that involves extracting a sub-section from the original image. This creates new images with a focus on specific areas or introduces variations in object size and position. There are three main cropping

Original Image   Horizontal Flip   Vertical Flip   Hor+Vert Flip

Fig. 1.3 Flipped augmented image samples created from one image

strategies commonly used: (a) Random Cropping: A random sub-region of the image is selected and extracted. This approach is simple to implement and helps the model learn to recognize objects even when they occupy only a part of the image frame. However, it can potentially remove important information from the edges if not done carefully. (b) Center Cropping: A square or rectangular region is extracted from the center of the image. This ensures the main object of interest remains in the cropped image but limits the variation in object position. It's useful when the object is centered in the original image and the background isn't crucial for classification. (c) Specific Cropping: Specific areas of interest are chosen for cropping, like the head in a portrait image. This provides more control over the information included in the cropped image and can be beneficial for tasks where focusing on specific parts is essential. However, it requires defining the cropping region beforehand, which can be time-consuming for large datasets. These help the model recognize the object even if only a part of it is visible. Figure 1.4 represents different cropped image samples.

1.3.1.5 Shearing: Shearing is a geometric image augmentation technique that alters the shape of an image to resemble a parallelogram. Imagine stretching one side of a rectangle upwards or downwards while keeping the other side fixed. This manipulation introduces a slanted perspective, similar to tilting the image slightly. During shearing, a specific pixel in the image is shifted proportionally based on its horizontal position. This controlled

Original Image   Random Crop   Center Crop   Specific Crop

Fig. 1.4 Cropped augmented image samples created from one image

Original Image   Vertical Shear   Horizontal Shear

Fig. 1.5 Sheared augmented image samples created from one image

distortion can be helpful for tasks like: (a) Simulating camera tilt: In real-world scenarios, cameras might not always be perfectly horizontal. Shearing can introduce variations that mimic slightly tilted camera angles, improving model robustness. (b) Accounting for perspective distortion: Objects in the distance often appear narrower due to perspective. Shearing can create variations that account for these perspective changes, helping the model recognize objects at different distances. (c) Enhancing specific features: In tasks like character recognition, a slight shear can be applied to emphasize the slant of letters, potentially improving model accuracy. However, it's important to control the degree of shearing to avoid introducing unrealistic distortions or losing important information from the image. Figure 1.5 represents different sheared image samples.

1.3.2 Color Space Transformations: Color space transformations encompass a range of image augmentation techniques that manipulate the way color information is represented in an image. This goes beyond simple adjustments like brightness and contrast. However, selecting the accurate transformation method depends on the specific task and the types of color variations that are relevant for achieving accurate model performance [4]. Here are some color space transformation techniques.

1.3.2.1 Brightness and Contrast Adjustment: Brightness and contrast adjustment are fundamental image augmentation techniques that manipulate the illumination properties of an image. This is crucial for training deep learning models in computer vision tasks because real-world images can be captured under varying lighting conditions. Here's a detail of their contributions: (a) Brightness Adjustment: This technique increases or decreases the overall intensity of pixels in the image, essentially simulating brighter or darker lighting scenarios. By introducing variations in brightness, the model learns to recognize objects even under different illumination conditions. This improves the model's generalizability and robustness to real-world lighting variations. (b) Contrast Adjustment: This technique alters the difference between the

Original Image   Brightness Adjustment   Contrast Adjustment

Fig. 1.6 Brightness/Contrast adjusted augmented image samples created from one image

lightest and darkest areas of the image. Increasing contrast enhances the distinction between objects and their background, potentially aiding in tasks like object detection. Conversely, decreasing contrast creates a more muted image, which can be helpful for tasks where subtle details might be important (like classifying different cloud formations). This helps the model perform well under different lighting scenarios. These techniques can be applied independently or combined to create even more diverse lighting variations. However, it's important to control the adjustments to avoid creating unrealistic lighting extremes or losing important details in the image. Figure 1.6 represents different brightness and contrast-adjusted image samples.

1.3.2.2 Color Jittering: Color jittering is a color space manipulation technique used for image augmentation in computer vision. It injects random variations into an image's color channels, effectively simulating slight changes in lighting or color balance that might occur in real-world scenarios. Here's how it benefits deep learning models: (a) Improved Generalizability: By introducing color variations, color jittering helps the model learn a more robust representation of the object's appearance. This reduces the risk of the model overfitting to the specific color distribution of the original training data. The model becomes more adaptable to recognizing objects even if their colors appear slightly different due to variations in lighting or camera settings. (b) Simulating Natural Color Variations: Natural objects can exhibit slight color variations due to factors like lighting, material properties, or even camera calibration. Color jittering introduces these subtle variations, helping the model become more robust to real-world color discrepancies. Color jittering typically involves adjusting three key color properties: hue, saturation, and brightness (value). The degree of jitter for each property is controlled by pre-defined ranges. This ensures the color variations remain realistic and don't drastically alter the overall appearance of the object. Overall, color jittering is a simple yet effective method for improving the color-related robustness of deep learning models in computer vision tasks. Figure 1.7 represents different color-jittered image samples.

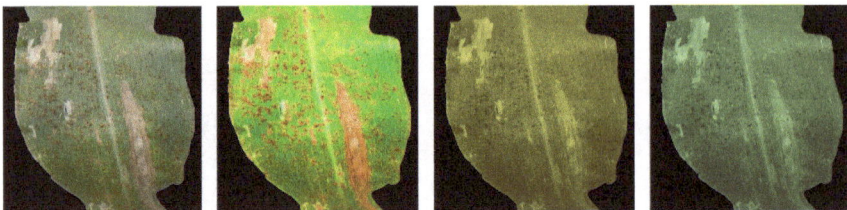

Fig. 1.7 Color-jittered augmented image samples created from one image

1.3.2.3 Grayscale Conversion: Grayscale conversion, while not as widely used as other augmentation techniques, can be a valuable tool in specific scenarios. It involves transforming an image from its original RGB format (containing red, green, and blue channels) to a single-channel format representing brightness intensity. Here's how it can benefit deep learning models: (a) Focus on Shape and Texture: By removing color information, grayscale conversion forces the model to learn and rely on features like object shape, texture, and spatial relationships between objects. This can be beneficial for tasks where color might be irrelevant or even misleading. For example, in a task like classifying different types of leaves, grayscale conversion might help the model focus on vein patterns and leaf shapes for accurate identification. (b) Reduced Model Complexity: Grayscale images require less processing power and memory compared to their RGB counterparts. This can be advantageous for training models on resource-constrained devices or for speeding up the training process, particularly when dealing with large datasets. However, grayscale conversion also has limitations. It discards valuable color information crucial for tasks where color plays a significant role. For instance, in tasks like classifying fruits or flowers, color is a key differentiating factor. Additionally, grayscale conversion might not be effective for tasks relying on specific color features, like traffic light detection. Therefore, grayscale conversion should be used strategically, considering the task at hand and the importance of color information for accurate classification or detection. Figure 1.8 represents the grayscale representation of the original image.

1.3.1 Other Traditional Techniques

1.3.3.1 Noise injection: Noise injection is a fundamental image augmentation technique that introduces random noise to an image. This simulates imperfections that might occur in real-world images due to factors like sensor noise, compression artifacts, or even environmental conditions. However, it's important to control the amount of noise injected. Excessive noise can overwhelm the actual image content, hindering the model's ability to learn

Original Image   Grayscale Image

Fig. 1.8 Grayscale-converted augmented image samples created from one image

the desired features [5]. The goal is to introduce controlled variations that mimic real-world noise, not to completely obscure the image information. Figure 1.9 represents the noisy representation of the original image.

1.3.3.2 Elastic Deformation: Elastic deformation is an image augmentation technique that introduces controlled distortions to an image, mimicking the warping and stretching that objects can undergo in real-world scenarios. Elastic deformation creates variations in object shape that traditional geometric transformations might not capture. This is particularly beneficial for tasks involving objects with non-rigid structures, like faces with expressions, fabrics with wrinkles, or even biological cells undergoing deformations. By encountering these variations during training, the model learns a more robust representation of the object class, improving its ability to generalize to unseen examples with slight shape variations. Unlike techniques like scaling or rotation that affect the entire image, elastic deformation allows

Fig. 1.9 Noisy augmented image samples created from one image

Fig. 1.10 Elastic deformed augmented image samples created from one image

for localized distortions. This can be particularly useful for tasks where focusing on specific regions of the object is important. For example, in facial recognition, elastic deformation can be applied around the eyes or mouth to simulate expressions, helping the model learn to identify faces even with slight variations in these features. However, elastic deformation requires careful parameter tuning. Excessive distortions can create unrealistic or unrecognizable images [6]. The goal is to introduce subtle variations that enhance model robustness, not to completely alter the object's appearance. Figure 1.10 represents the elastic deformed version of the original image.

There are some benefits of using traditional image augmentation techniques. These techniques are relatively easy to implement using existing image-processing libraries. They require minimal computational resources compared to more advanced methods. Traditional techniques are effective in improving the performance of deep learning models for various computer vision tasks.

Traditional image augmentation techniques, while effective, have limitations that hinder their ability to fully address real-world data complexities [7]. These limitations stem from their predefined nature. Firstly, they might not capture the full range of variations present in real-world scenarios. Rotations by fixed angles or scaling by specific factors might not encompass the continuous spectrum of transformations an object can undergo. Secondly, the lack of control over the augmentation process can lead to unrealistic or unnatural variations. Random cropping might remove crucial information from the image, while fixed-value color jittering might introduce nonsensical color combinations. Thirdly, traditional techniques often struggle with complex deformations or occlusions. Applying elastic deformations with a single set of parameters might not generalize well to the vast range of facial expressions or object interactions encountered in real-world images. Finally, these techniques don't inherently introduce new information or concepts not present in the original dataset. This limits their ability to improve model performance on entirely unseen scenarios. These limitations necessitate exploring more advanced augmentation methods that can learn and adapt to the specific dataset and task at hand.

1.4 Deep Learning for Image Augmentation

While traditional image augmentation techniques have served as a workhorse for computer vision tasks, deep learning-based augmentation offers significant advantages that push the boundaries of performance and generalizability [8]. Here's a detail of why deep learning methods are becoming increasingly crucial: (a) Limited Control and Coverage: Traditional techniques like rotation or scaling offer limited control over the variations introduced. They might not capture the full spectrum of complexities present in real-world data (like subtle facial expressions or object interactions). Deep learning, on the other hand, can learn intricate relationships within the data. Techniques like Generative Adversarial Networks (GANs) can produce highly realistic and diverse image variations that go beyond simple geometric transformations. This allows for a more comprehensive exploration of the data distribution, enhancing model robustness. (b) Inability to Adapt to Specific Tasks: Traditional techniques are generic and don't inherently adapt to the specific needs of a task. For instance, random cropping might remove crucial information for object detection tasks. Deep learning methods can be tailored to the task at hand. Variational Autoencoders (VAEs) allow for targeted augmentation by manipulating the latent space, focusing on generating variations in specific aspects of the image relevant to the task (like object pose in human pose estimation). (c) Limited Realism and Generalizability: Traditional techniques often introduce unrealistic distortions or unnatural color variations. These can hinder model performance on unseen data that deviates slightly from the augmented variations. Deep learning approaches like GANs can create highly realistic images that closely resemble real-world scenarios. This bridges the gap between training data and real-world deployment, improving the model's capability to generalize to unseen variations and improve overall performance. (d) Lack of Automation and Efficiency: Finding the optimal combination of traditional augmentation techniques for a specific task can be a time-consuming process of trial and error. Deep learning methods like AutoAugment utilize reinforcement learning to automatically discover the most effective augmentation strategies. This saves time and effort in the data preparation process while ensuring the chosen augmentations are most beneficial for the specific task.

While traditional techniques have their place, deep learning-based image augmentation offers a more sophisticated and powerful approach. By using the capabilities of neural networks, these methods can create highly realistic, task-specific variations that significantly improve the performance and generalizability of deep learning models in computer vision tasks. As research in this area continues to evolve, deep learning augmentation is poised to play an even greater role in future of computer vision. Here's a glimpse into this exciting area.

1.4.1 Generative Adversarial Networks (GANs): This approach [9] pits two neural networks against each other. One network (generator) tries to create new, realistic images based on the training data, while the other network (discriminator) attempts to distinguish between real and generated images. This adversarial training process pushes the generator to create increasingly realistic and diverse image variations that can significantly expand the training dataset.

1.4.2 Variational Autoencoders (VAEs): VAEs [10] are a type of neural network that learns a compressed representation of the training data. This compressed representation, called the latent space, captures the essential features and variations within the data. By manipulating points in the latent space, VAEs can produce new images that follow the learned data distribution. This allows for targeted augmentation, where specific aspects of the image (like object pose or lighting) can be controlled during the generation process.

1.4.3 AutoAugment: This technique utilizes reinforcement learning to automatically discover the most effective image augmentation policies for a specific task and dataset [11]. It explores different combinations of traditional and deep-learning-based augmentation techniques, evaluating their impact on model performance. This allows for an automated approach to finding the optimal augmentation strategy, saving time and effort in the data preparation process.

Deep learning-based image augmentation offers a powerful way to not only expand datasets but also steer the augmentation process toward generating images that are most beneficial for the specific task at hand. This opens doors for significant improvements in model performance and generalizability in computer vision tasks.

1.5 Summary

Limited image data poses a significant challenge for deep learning models in computer vision tasks. This chapter delved into traditional image augmentation techniques and explored how deep learning advancements are pushing the boundaries of data augmentation. The chapter began by outlining the core challenges of limited data: overfitting, poor generalizability, difficulty learning complex concepts, and increased training time/cost. It then explored various traditional image augmentation techniques that address these challenges by artificially expanding the dataset: (a) Geometric Transformations: Rotations, scaling, flipping, and cropping introduce variations in object size, position, and orientation. (b) Color Space Transformations: Brightness/contrast adjustments and color jittering manipulate illumination and color properties, simulating real-world variations. (c) Other Techniques: Noise injection and elastic deformation introduce imperfections and controlled distortions, enhancing model robustness. While effective, traditional techniques have limitations: (a) Limited Variation: They might not capture the full range of real-world complexities. (b) Lack of Control: Variations introduced might be unrealistic or not specific to the task. (c) Limited Realism: Traditional methods often struggle with complex deformations or occlusions. (d) Inability to Introduce New Information: They don't inherently create entirely new concepts not present in the original data. The chapter then introduced deep learning-based image augmentation techniques that overcome these limitations. The chapter concluded by highlighting the key advantages of deep learning image augmentation: (a) Superior Control and Coverage: Deep learning methods can capture intricate data relationships and generate variations beyond simple geometric transformations. (b) Adaptability to Specific Tasks: Techniques like VAEs allow for targeted augmentation, focusing on variations relevant to the task at hand. (c) Enhanced Realism and Generalizability: Deep learning approaches create highly realistic images, improving model performance on unseen data variations. (d) Automation and Efficiency: Techniques like AutoAugment automate the process of discovering optimal augmentation strategies, saving time and effort. Deep learning-based image augmentation represents a significant leap forward in addressing the limitations of traditional techniques. As research continues, these methods are poised to play an even greater role in future of computer vision, enabling the development of more robust, generalizable, and high-performing deep learning models.

References

1. Mittal, S. M., Sung, A. H., & Tamersoy, S. (2021, December) The many shades of limited data in computer vision. In *IEEE transactions on pattern analysis and machine intelligence* (vol. 43, no. 12, pp. 4075–4088).
2. Xie, S., Zheng, H., Chen, C., & Zhao, J. (2021). A survey on image data augmentation for deep learning. *IEEE Transactions on Pattern Analysis and Machine Intelligence, 43*(4), 1241–1256.
3. Tran, L., Yin, X., & Liu, X. (2021). A survey of geometric data augmentation for robust deep learning. *IEEE Transactions on Neural Networks and Learning Systems, 32*(11), 5237–5252.
4. Zhang, H., Xu, Y., Ye, X., & Li, X. (2021). A comprehensive survey on color space transformations for image augmentation in deep learning. *IEEE Transactions on Image Processing, 30*, 3850–3865.
5. Han, S., Lee, H., & Kim, J. (2022). Noise injection: A survey on applications and techniques in deep learning. *IEEE Transactions on Neural Networks and Learning Systems, 33*(4), 1375–1390.
6. Gupta, S. S., & Maheshwari, A. (2022). Elastic deformation augmentation techniques for deep learning: A comprehensive review. *IEEE Transactions on Image Processing, 31*, 1234–1248.
7. Zhang, J., Li, H., Wang, Y., & Yang, Q. (2022). Beyond traditional image augmentation: A survey on addressing real-world data complexities. *IEEE Transactions on Pattern Analysis and Machine Intelligence, 44*(7), 1913–1929.
8. Gupta, S. J., & Dhall, A. (2022). Deep learning-based image augmentation techniques: A comprehensive survey. *IEEE Transactions on Image Processing, 31*, 4567–4583.
9. Zhang, J., Wang, X., Li, Y., & Wu, Q. (2022). Generative adversarial networks for image augmentation: A comprehensive review. *IEEE Transactions on Pattern Analysis and Machine Intelligence, 44*(9), 2817–2834.
10. Liu, Y., Zhang, H., Wang, J., & Li, L. (2022). Variational autoencoders for targeted image augmentation: A review. *IEEE Transactions on Image Processing, 31*, 789–802.

11. Cubuk, D., Zoph, D., Shlens, J., & Le, Q. V. (2021). Rethinking data augmentation for image super-resolution with autoaugment. *IEEE Conference on Computer Vision and Pattern Recognition (CVPR)*, June 2021.

Chapter 2
Generative Adversarial Networks Based Image Augmentation

Generative Adversarial Networks (GANs) have emerged as a powerful tool for image augmentation, offering a more sophisticated approach to expanding training datasets. Unlike traditional methods that rely on simple transformations like rotations and flips, GAN-based augmentation uses the generative capabilities of these networks to synthesize highly realistic and diverse variations of existing images. This chapter will explore the principles and applications of GAN-based image augmentation, highlighting its potential to significantly improve the performance and robustness of deep learning models in computer vision.

2.1 Introduction to GANs

GANs denote a revolutionary progression in the field of generative modeling, enabling the creation of remarkably realistic synthetic data. Unlike traditional generative models that primarily rely on maximum likelihood estimation, GANs introduce a competitive framework that drives the generation of highly convincing outputs. GANs are a class of machine learning frameworks designed by Ian Goodfellow and his colleagues in 2014 [1]. They pit two neural networks against each other in a competitive process to generate new data instances that resemble the training data.

At the core of a GAN lies a two-player game between two neural networks: the generator and the discriminator.

J. Chaki, *The Art of Deep Learning Image Augmentation: The Seeds of Success*, SpringerBriefs in Computational Intelligence, https://doi.org/10.1007/978-981-96-5081-1_2

2.1.1 Generator

The generator is a neural network responsible for creating new data instances. Its architecture typically consists of several layers, such as convolutional, deconvolutional, and fully connected layers [2]. The generator in a GAN is essentially a complex function that maps a random noise vector to a data point in the desired output space. Mathematically, we can represent this as shown in Eq. (2.1).

$$G(z) = x_{\text{fake}}, \tag{2.1}$$

where: G is the generator function, z is the random noise vector drawn from a prior distribution (usually a Gaussian or uniform distribution) and x_{fake} is the generated data point. A typical generator architecture for image generation consists of multiple layers that progressively build up a high-resolution image from a random noise vector. The core components include:

Input Layer: The input layer of a generator is the foundational component that sets the stage for the entire image generation process. Noise injection at the input layer of a generator involves adding random noise to the initial random noise vector before it's fed into the network. This technique is a form of regularization that helps prevent the generator from overfitting to the training data and encourages it to explore a wider range of possible outputs. By introducing randomness at the very beginning of the generation process, the generator is forced to learn more robust and generalizable features. The most common noise distribution used for the input layer is the standard normal distribution (Gaussian distribution with mean 0 and variance 1). This choice is motivated by its mathematical properties and its capability to capture a wide range of variations. The mean of zero ensures that the noise is centered around zero, and the unit variance provides a suitable scale for the generator to operate. It can be easily scaled and transformed to fit different generator architectures. Mathematically, the noise vector z can be represented as shown in Eq. (2.2).

$$z \sim N(0, I), \tag{2.2}$$

where: z is the noise vector, $N(0, I)$ signifies a multivariate normal distribution with identity covariance matrix and zero mean.

While the standard normal distribution is a solid choice, other distributions can be explored to introduce specific properties into the generated images. Uniform distribution can be used to explore a wider range of values, potentially leading to more diverse outputs. However, it might lack the smoothness and continuity associated with the normal distribution. Laplacian distribution has heavier tails compared to the normal distribution, which can potentially introduce more extreme values in the generated images. It might be suitable for generating images with sharp contrasts or outliers.

The choice of noise distribution can significantly impact the generator's ability to learn complex patterns and generate high-quality images. A well-suited noise distribution can improve convergence speed, improve the diversity of produced images, and facilitate the capture of specific image characteristics. The noise vector's dimension, often denoted as the latent space dimension, is a crucial hyperparameter that significantly affects the generator's capacity to generate diverse and high-quality images. A smaller latent space limits the generator's ability to produce complex and varied images. It might lead to mode collapse, where the generator produces similar outputs. A larger latent space provides more degrees of freedom for the generator, allowing it to capture intricate details and generate a wider range of images. However, it also increases the complexity of the training process and can lead to overfitting. Experimentation is crucial to determine the optimal noise distribution, optimal latent space dimension for a given dataset, and generator architecture.

To further enhance the generator's capabilities, various techniques can be applied to manipulate the noise vector. Conditional noise incorporating additional information, such as class labels or attributes, into the noise vector can guide the generator to generate specific types of images. This is commonly used in conditional GANs. There are several ways to incorporate conditioning information into the generator. The conditioning information can be concatenated with the noise vector before being fed into the generator. The conditioning information can be used as parameters for batch normalization layers in the generator. Attention mechanisms can be used to focus the generator's attention on specific parts of the conditioning information. Hierarchical Noise is a technique employed in GANs to introduce a multi-scale approach to image generation. This method involves structuring the input noise into multiple levels, each contributing to different aspects of the generated image. By breaking down the noise into hierarchical levels, the generator can progressively build up the image from coarse to fine details. This approach enhances the generator's ability to capture intricate patterns and generate highly realistic and diverse images. The core idea is to create a pyramid-like structure of noise vectors, where the top level represents the most abstract features of the image, and subsequent levels introduce progressively finer details. This hierarchical organization allows the generator to focus on specific image components at each stage, leading to improved control over the generation process. By incorporating hierarchical noise, GANs can generate images with more complex structures and greater realism, as the generator can effectively learn to map different levels of noise to corresponding image features. This technique has been instrumental in achieving state-of-the-art results in various image-generation tasks.

Noise injection at the input layer can be particularly effective in preventing mode collapse, a common issue in GAN training where the generator produces only a limited set of similar outputs. By introducing random perturbations to the input noise, the generator is encouraged to explore different regions of the latent space, leading to a more diverse set of generated images. While noise injection at the input layer is a straightforward technique, its effectiveness can vary depending on the complication of the generator architecture and the nature of the training data. It's often combined with other regularization methods, like weight decay or dropout, to achieve optimal results.

The input layer acts as a bridge between the random space and the image space. The generator's subsequent layers are tasked with transforming this random noise into a meaningful image representation. By introducing randomness at the input layer, the generator can produce a variety of different images, even when presented with the same noise vector during different training iterations. In essence, the input layer is the starting point for the generator's creative process, providing the raw material from which the network can build intricate and realistic images.

For example, for generating human faces, a combination of Gaussian noise and conditional noise based on gender and age can be used. A high-dimensional noise vector can provide sufficient variability for different facial features, while the conditional information ensures the generated images adhere to the specified attributes.

Convolutional Layers: Convolutional layers are fundamental building blocks in the generator of a GAN, especially for image generation tasks. They are responsible for extracting and manipulating features from the input noise vector, gradually constructing the desired image output. A convolutional layer applies a set of learnable filters to the input data, producing feature maps. Each filter slides over the input, performing element-wise multiplications and summing the results. The size of the convolutional filters determines the receptive field of the layer. Smaller filters (e.g., 3×3) are commonly used to capture local patterns, while larger filters can capture more global information. The stride parameter controls the movement of the filter over the input. A stride of 1 is often used to preserve spatial information, while larger strides can downsample the feature maps. Padding can be utilized to preserve the spatial dimensions of the output feature maps. Zero-padding is commonly used to maintain spatial dimensions after convolution. The number of filters determines the dimensionality of the output feature maps.

The output of a convolutional layer can be calculated as represented in Eq. (2.3).

$$y[i,j] = f\left(\text{sum}\left(x[i+k, j+l] \times w[k, l]\right) + b\right), \qquad (2.3)$$

where: $y[i,j]$ is the output at pixel (i,j), $x[i+k, j+l]$ is the input at pixel $(i+k, j+l)$, $w[k, l]$ are the weights of the filter, b is the bias, f is the activation function (e.g., ReLU).

A convolutional layer generates a collection of feature maps, where each map highlights a distinct characteristic identified within the input data. These feature maps capture essential patterns and structures within the data, like corners, edges, and textures. As the network goes deeper, the convolutional layers extract increasingly complex features. Early layers might detect simple patterns like edges and corners, while later layers combine these low-level features to form more abstract representations, such as shapes and objects. The weights of the convolutional filters are learned during the training process. The generator aims to produce images that can fool the discriminator, and the discriminator provides feedback through the gradient backpropagation process. This iterative procedure permits the generator to enhance

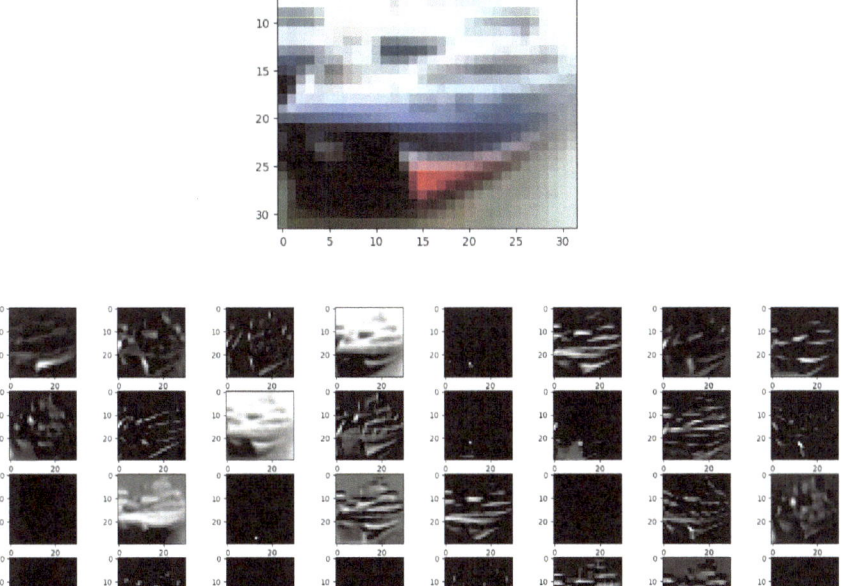

Fig. 2.1 Feature maps generated by convolutional layers

its feature extraction capabilities. Figure 2.1 represents 32 generated feature maps
by using the convolutional filters. This experimentation was conducted on CIFAR10
dataset samples.

Once the generator has extracted meaningful features from the input noise
vector, it's essential to gradually increase the spatial dimensions of these features to
reconstruct a high-resolution image. This is where transposed convolutional layers,
frequently denoted as deconvolutional layers, come into play. Mathematically, a
transposed convolution can be seen as the inverse operation of a standard convolution. Though, it's important to note that it's not a true inverse, as the goal is not to
perfectly reconstruct the original input but to generate new data. The process involves
upsampling and convolution. Upsampling means to increase the spatial dimensions
of the input feature map by inserting zeros between pixels and during convolution
a convolutional filter is applied to the upsampled feature map. Figure 2.2 shows an
example of a transposed convolutional operation.

Transposed convolutions can sometimes produce checkerboard patterns in the
output. Techniques like fractional-strided convolutions or sub-pixel convolutions can
help mitigate this issue. The core idea of fractional-strided convolutions is to apply
a standard convolution to the input feature map. Then the output of the convolution
is rearranged into a higher-resolution feature map by interleaving pixel values.

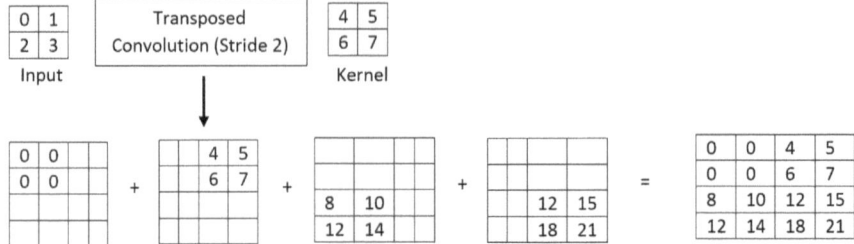

Fig. 2.2 Transposed convolution operation

Batch Normalization: Batch normalization normalizes the activations of a layer across a batch of training examples. It addresses the problem of "Internal Covariate Shift," where the distribution of inputs to a layer changes during training as the parameters of previous layers change. The process involves normalizing the activations of a layer for each mini-batch of data. This is achieved by deducting the batch mean and dividing it by the batch standard deviation. The normalized activations are then scaled and shifted using learnable parameters, gamma and beta, respectively. Mathematically, the batch normalization process can be represented as follows by using Eq. (2.4).

$$y = \gamma \times \frac{(x - \mu)}{\sqrt{\text{var} + \varepsilon}} + \beta, \qquad (2.4)$$

where: y is the normalized output, x is the input, μ, and var are the mean and variance of the batch, ε is a small constant to prevent division by zero, γ and β are learnable parameters.

By normalizing the activations, batch normalization helps to stabilize the training procedure, reduce the vanishing/exploding gradient issue, and improve the generalization ability of the model. Additionally, it allows for the use of higher learning rates without compromising stability.

Activation Functions: Nonlinear activation functions are introduced to introduce nonlinearity into the network. Common choices include ReLU: $f(x) = \max(0, x)$, LeakyReLU: $f(x) = \max(\alpha \times x, x)$, Tanh: $f(x) = (\exp(x) - \exp(-x))/(\exp(x) + \exp(-x))$

Output Layer: The output layer of a GAN generator is the final stage in the image-generation process. Its primary function is to transform the high-dimensional feature representation produced by the preceding layers into a visually interpretable image format. A common choice for the activation function in the output layer of image-generation tasks is the hyperbolic tangent (tanh) function. This function scales the output values to a range between -1 and 1, which corresponds to the pixel intensity values typically used in image representation. This ensures that the generated images have appropriate contrast and dynamic range. While tanh is commonly used, other activation functions like sigmoid can also be employed. However, tanh often provides better results in terms of image quality. The output layer's dimensions should match the desired image size. For example, to generate a 64 × 64 pixel image with 3 color channels, the output layer should have dimensions of 64 × 64 × 3. The generator's loss function, typically adversarial loss, drives the learning process and indirectly influences the output layer's behavior.

By carefully designing the architecture and hyperparameters, the generator can learn to offer highly realistic images.

The generator's objective is to minimize the discriminator's ability to correctly classify its generated samples as fake. This can be mathematically expressed by using Eq. (2.5).

$$\min_G\left[\log(1 - D(G(z)))\right], \tag{2.5}$$

where: $\min_G$ denotes minimizing the loss concerning the generator's parameters, E is the expectation over the random noise vector z, D is the discriminator function.

However, this loss function can lead to vanishing gradients during the initial training stages. To overcome this, a different loss function is often used as shown in Eq. (2.6).

$$\max_G[\log(D(G(z)))] \tag{2.6}$$

2.1.1.1 Discriminator

The discriminator serves as a fundamental part of a GAN, acting as a gatekeeper that differentiates between real and fake data. In the context of GANs, "real data" typically refers to genuine samples from the training dataset, while "generated data" pertains to outputs produced by the generator, which aims to replicate the characteristics of real samples [3]. This binary classification task is crucial since the discriminator's performance can significantly influence the efficacy and success of the GAN's training procedure.

The discriminator in a GAN is primarily composed of multiple layers, starting with convolutional layers that function to extract meaningful features from the input data. Convolutional layers apply filters to the data, helping to identify patterns such as edges, textures, and shapes, which are crucial for distinguishing between genuine samples and those produced by the generator. Following these layers, fully connected layers consolidate the extracted features and process them to generate a final output, typically a scalar value on behalf of the probability that the input is a "real" sample from the dataset rather than a "fake" generated one.

The design of the discriminator can differ significantly reliant on the nature of the input data. For instance, images present different challenges and characteristics that require tailored architectures to effectively extract relevant features. In the case of image data, the discriminator may utilize deeper architectures with more convolutional layers to capture intricate visual details.

This flexibility in architecture is key to the discriminator's ability to perform well across different types of data. By customizing the structure and incorporating domain-specific feature extraction mechanisms, the discriminator can effectively learn the subtle differences between real and generated samples. Such adaptability is beneficial as it allows for the application of GANs in a wide range of domains, from image generation to even complex tasks such as image-to-image translation.

In a GAN framework, the training process involves both the discriminator and the generator being updated concurrently. This simultaneous optimization is essential as it establishes a dynamic interaction between the two components. The discriminator continuously refines its ability to classify samples, while the generator adapts in response to the discriminator's evaluations. This interplay is fundamental to the adversarial learning paradigm that defines GANs.

The primary objective of the discriminator is twofold: it seeks to maximize its performance in identifying real samples correctly and minimize the probability of mistakenly accepting generated samples as real. These dual aims mean that the discriminator becomes increasingly adept at recognizing the details and features that differentiate real data from synthetic data. Therefore, the discriminator uses loss functions, often based on binary cross-entropy, to guide its learning in this regard.

As the discriminator continues its training, it becomes increasingly skilled at identifying subtle and complex features that differentiate real data from synthetic data. This skill development is crucial because the generator iteratively improves its outputs based on the feedback provided by the discriminator. The discriminator's enhanced capability leads to more nuanced feedback, allowing the generator to tweak its outputs more effectively. Consequently, this feedback loop is fundamental for elevating the excellence of generated instances. To guide its learning process, the discriminator employs loss functions that quantify its performance. A commonly used function is the binary cross-entropy loss, which measures the difference between the predicted probabilities of the discriminator and the actual labels (real or fake). This function supports the adversarial training process by penalizing misclassifications, thereby encouraging the discriminator to continually refine its model. In essence, the loss function acts as a compass, steering the training toward better performance metrics. The discriminator's training incorporates optimization algorithms that adjust its weights based on the calculated loss. These algorithms, such as Stochastic Gradient Descent (SGD) or its variants like Adam, help minimize the loss function over numerous iterations. As the discriminator learns to minimize its loss through these adjustments, it becomes adept at classifying data, ultimately fortifying the GAN's ability to generate realistic outputs.

The adversarial nature of GANs can be conceptualized as a zero-sum game, where the gain of one network results in a loss for the other. In this context, when the generator improves its ability to generate realistic samples, it creates a trickier challenge for the discriminator to differentiate between the two. Conversely, if the discriminator becomes too effective, the generator may struggle to produce outputs that the discriminator cannot recognize as artificial. This competitive balancing creates a robust training environment that fosters continual improvement in both networks.

The adversarial setup facilitates a vital feedback mechanism where the performance of the discriminator directly influences the behavior of the generator. As the discriminator gets better at spotting fakes, the generator receives signals to improve the quality and realism of its outputs. This iterative procedure confirms that both components evolve together, leading to more sophisticated generated data as training progresses. Ultimately, this feedback loop is crucial for achieving high-quality results from GANs.

While this adversarial interaction is beneficial for fostering quality improvements, it can also introduce training instability. Overly powerful discriminators can lead to situations where the generator fails to learn adequately, resulting in issues like mode collapse or the generator producing limited variations of outputs. Consequently, maintaining a balance in the learning capacity of both the discriminator and the generator is vital for ensuring effective training and convergence toward optimal outputs.

The progress of the discriminator directly influences the generator's capability to generate superior instances. If the discriminator effectively maximizes its performance in distinguishing real from fake, the generator is compelled to innovate and improve its data generation strategies. This interdependent relationship not only propels the capabilities of both networks but also underlines the essence of the adversarial setup within GANs.

The generator and discriminator are trained in an adversarial manner. The generator's objective is to maximize the probability of the discriminator making a mistake, while the discriminator aims to minimize the probability of being fooled by the generator. This competitive process drives both networks to improve their performance over time. Figure 2.3 illustrates the generator producing fake data, the discriminator classifying data as real or fake, and the backpropagation updates to both networks. Figure 2.3 represents the working principle of GAN.

Figure 2.4 represents the adversarial training process of a GAN.

A PatchGAN discriminator [4] is a variant of the traditional GAN discriminator that operates on image patches rather than the entire image. Instead of outputting a single scalar value indicating whether the entire image is real or fake, the PatchGAN outputs a map of real/fake decisions for overlapping patches of the image. It focuses on local image patches rather than the global image structure. It typically employs convolutional layers to process the image and produce patch-wise decisions. It generates an output map where each pixel represents the probability of the corresponding image patch being real. The final decision is often made by averaging the probabilities

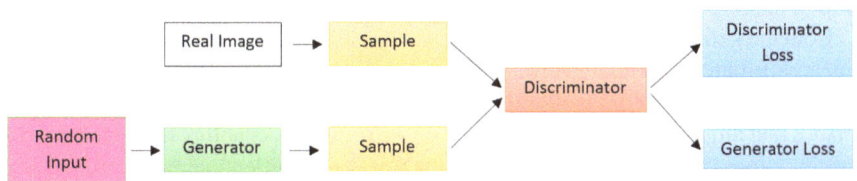

Fig. 2.3 The working principle of GAN

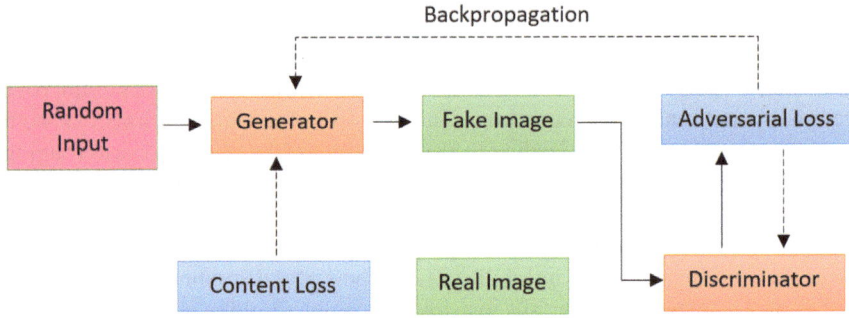

Fig. 2.4 The adversarial training process of a GAN

from all patches. By focusing on local patches, PatchGAN can capture finer details and textures in the image. Compared to full-image discriminators, PatchGAN can be more efficient for larger images. Less sensitive to global image statistics, making it more robust to variations in image content. The size of the patch can significantly impact the discriminator's performance. A larger patch size allows for capturing more context, but it also increases computational cost. A common choice is a patch size of 70 × 70 pixels. To improve the discriminator's ability to detect fine-grained details, overlapping patches can be used. This helps to capture more information about the image content. The discriminator's output is a feature map with the same spatial dimensions as the input patch, where an individual pixel denotes the probability of the corresponding patch being real. This permits the generator to learn from local feedback, improving the quality of generated images. By operating on image patches, PatchGAN effectively captures local image structures and helps the generator produce more realistic and detailed outputs. Figure 2.5 represents the patchGAN discriminator. Each value of the prediction matrix represents whether the image patch is real or artificially generated.

2.2 Conditional GAN

Conditional GANs (cGANs) are an extension of standard GANs that introduce additional information, known as conditioning information, to both the generator and discriminator [5]. This conditioning allows for more control over the generated output, making them particularly suitable for image augmentation tasks.

Both the generator and discriminator receive additional information, such as class labels, attributes, or other relevant data, along with the standard inputs (random noise for the generator, image for the discriminator). The generator learns to produce images that align with the given conditioning information, enabling the generation of specific types of images. The discriminator learns to differentiate between real and

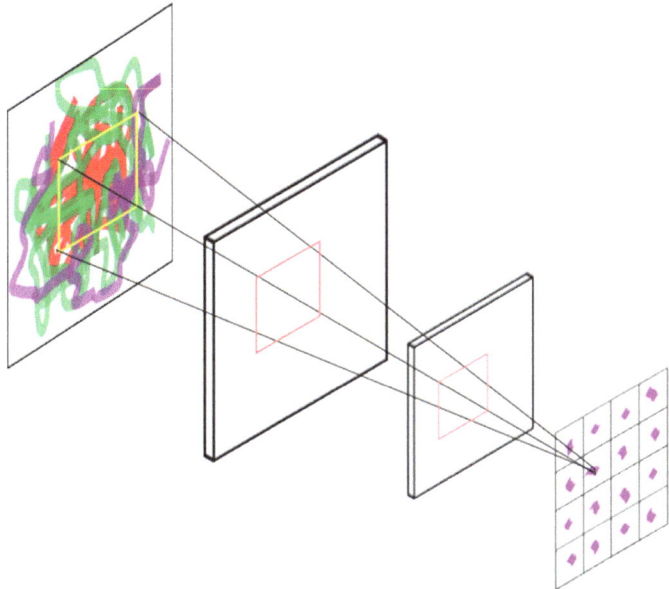

Fig. 2.5 PatchGAN discriminator

fake images while also considering the conditioning information, making it more robust.

The process involved in cGAN for image generation is as follows. First, collect a dataset with images and corresponding labels or attributes. Then a generator and discriminator architecture is created that incorporates the conditioning information. The generator in a cGAN takes both a random noise vector, z, and conditioning information, c, as input. The conditioning information can be in various forms, such as class labels, attributes, or other relevant data. The simplest method is to concatenate the noise vector and the conditioning information into a single vector. This combined vector is then fed into the generator network.

$$z_{combined} = \text{concat}(z, c)$$
$$G(z_{combined}) = x_{fake}, \tag{2.7}$$

where: $z_{combined}$ is the concatenated vector of noise and conditioning information, G is the generator function, x_{fake} is the generated image.

Instead of concatenating, the conditioning information can be used as parameters for batch normalization layers within the generator. This allows for more flexible control over the generation process.

The discriminator in a cGAN takes both the input image, x, and the conditioning information, c, as input. It aims to classify the input as real or fake, considering the given condition.

$$x_{\text{combined}} = \text{concat}(x, c) \tag{2.8}$$

$$D(x_{\text{combined}}) = \text{probability of } x \text{ being real given } c,$$

where: x_combined is the concatenated image and conditioning information, D is the discriminator function.

Another approach is to provide the conditioning information as an additional input to the discriminator. This allows the discriminator to focus on different features based on the condition.

The core of GAN training is the adversarial loss function, which pits the generator against the discriminator. The discriminator aims to maximize the probability of correctly classifying real and fake images. This can be formulated as shown in Eq. (2.9).

$$L_D = -\big[\log(D(x|c))\big] - \big[\log(1 - D(G(z, c)|c))\big], \tag{2.9}$$

where: L_D is the discriminator loss, $D(x|c)$ is the probability that the discriminator assigns to a real image x given the condition c, $D(G(z, c)|c)$ is the probability that the discriminator assigns to a fake image $G(z, c)$ given the condition c, E is the expectation over the real data distribution and the noise distribution z.

The generator aims to minimize the probability of the discriminator correctly classifying its generated images as fake. This can be formulated as represented in Eq. (2.10).

$$L_G = -\big[\log(D(G(z, c)|c))\big], \tag{2.10}$$

where: L_G is the generator loss.

The training process involves an iterative update of the generator and discriminator. Draw a batch of real images and corresponding conditions from the training set. Generate a batch of fake images using the generator with random noise and the same conditions. Calculate the discriminator loss for both real and fake images. Update the discriminator's parameters using backpropagation to minimize the loss. Generate a new batch of fake images. Calculate the generator loss based on the discriminator's output. Update the generator's parameters using backpropagation to maximize the generator loss.

Once a cGAN has undergone rigorous training, it becomes a versatile tool for creating new images based on specific conditions. To generate an image, a desired condition is first defined, such as a class label, attribute, or other relevant information. This condition is then encoded into a suitable format compatible with the generator's input. Subsequently, a random noise vector is generated, serving as a foundation for the image. This noise vector, combined with the encoded condition, is fed into the trained generator model. The generator, having learned the complex mapping between noise, conditions, and images during training, produces a new image that aligns with the specified condition. The diversity and quality of the generated images depend on factors such as the complexity of the cGAN architecture, the quality of the training data, and the precision of the conditioning information.

Pix2Pix is a specific implementation of a cGAN designed for image-to-image translation tasks [6]. It differs from traditional GANs by incorporating input image information into both the generator and discriminator. This conditioning allows the model to learn a direct mapping between input and output images. Here the generator is A U-Net architecture is commonly used for the generator, which enables efficient learning of complex mappings between input and output images. The generator takes the input image as input and produces the corresponding output image. A PatchGAN discriminator is often employed, which classifies image patches as real or fake. This approach encourages the generator to produce realistic details at multiple scales. The loss function combines adversarial loss, L1 loss (as a content loss), and optionally a gradient loss to preserve image structures. Pix2Pix learns a direct mapping between input and output images. It often produces high-quality results with preserved details. It can be applied to various image-to-image translation tasks. By using the conditional nature of GANs and incorporating the U-Net architecture, Pix2Pix has achieved impressive results in image-to-image translation tasks.

2.3 CycleGAN

CycleGAN is a type of GAN that does not require paired training data. It is particularly useful for image-to-image translation tasks where paired data is scarce or unavailable [7]. A CycleGAN consists of two generators and two discriminators. The generator G in a CycleGAN is responsible for mapping images from domain X (e.g., photos) to domain Y (e.g., paintings). It learns to capture the underlying style and structure of domain Y and apply it to the input image from domain X. This process involves complex transformations, including changes in color palette, texture, and overall visual appearance. The generator's architecture typically consists of convolutional and deconvolutional layers, along with activation functions and normalization techniques. The generator F in a CycleGAN operates in the opposite direction to the generator G. Its primary function is to translate images from domain Y (e.g., paintings) back to domain X (e.g., photos). This reverse translation is crucial for enforcing cycle consistency. By reconstructing the original image after undergoing a round-trip translation, the CycleGAN ensures that the generated images maintain semantic and structural integrity. Similar to generator G, generator F typically employs convolutional and deconvolutional layers to capture and manipulate image features. The goal of F is to learn the inverse mapping of the style transformation introduced by generator G, bringing the image back to its original domain. The discriminator Dx in a CycleGAN acts as a binary classifier, tasked with distinguishing between real images from domain X and fake images generated by the generator G. It operates on a similar principle to traditional image classification networks, employing convolutional layers to extract appropriate features. The discriminator aims to maximize its accuracy in classifying input images as real or fake. By doing so, it provides essential feedback to the generator, encouraging it to produce more realistic and convincing outputs. The discriminator Dy in a CycleGAN acts as a binary classifier, tasked with distinguishing between real images from domain Y (e.g., paintings) and fake images generated by the generator F. Similar to Dx, it employs convolutional layers to extract relevant features from the input image. The discriminator Dy aims to maximize its accuracy in classifying images as real or fake. By doing so, it provides essential feedback to the generator F, encouraging it to produce more realistic and convincing outputs in domain Y. The loss function for Dy is similar to that of Dx, measuring the binary cross-entropy between the predicted probability and the ground truth labels. This adversarial process between Dy and F drives the improvement of the generator F in producing high-quality images in domain Y.

2.3.1 Loss Functions

The CycleGAN loss function comprises several components. The details are as follows.

2.3.1.1 Adversarial Loss in CycleGAN

The adversarial loss in CycleGAN is a key component that drives the competition between the generator and discriminator [8]. It encourages the generator to produce highly realistic images that can fool the discriminator, while the discriminator aims to accurately distinguish between real and fake images.

The discriminator's goal is to maximize the probability of correctly classifying real images as real and fake images as fake. This can be formulated using a binary cross-entropy loss.

For discriminator Dx:

$$L_{D_X} = -\big[\log(D_X(x))\big] - \big[\log(1 - D_X(G(y)))\big] \tag{2.11}$$

For discriminator Dy:

$$L_{D_y} = -\big[\log(D_Y(y))\big] - \big[\log(1 - D_Y(F(x)))\big], \tag{2.12}$$

where: $D_X(x)$ is the probability that the discriminator D_X assigns to a real image x from domain X being real, $D_X(G(y))$ is the probability that the discriminator D_X assigns to a fake image $G(y)$ from domain X being real, $D_Y(y)$ is the probability that the discriminator D_Y assigns to a real image y from domain Y being real, $D_Y(F(x))$ is the probability that the discriminator D_Y assigns to a fake image $F(x)$ from domain Y being real, E is the expectation over the real data distribution.

The generator's goal is to minimize the probability of the discriminator correctly classifying its generated images as fake. This can be formulated as follows.

For generator G:

$$L_{GX} = -\left[\log(D_X(G(y)))\right] \tag{2.13}$$

For generator F:

$$L_{GY} = -\left[\log(D_Y(F(x)))\right] \tag{2.14}$$

The adversarial loss encourages a competitive relationship between the generator and discriminator, driving both models to improve their performance over time.

2.3.1.2 Cycle Consistency Loss

Cycle Consistency Loss (CCL) is a vital component of the CycleGAN framework, ensuring that the generated images maintain semantic and structural consistency with the original images [9]. The core idea behind cycle consistency is to enforce a cyclic mapping between the two domains. If an image from domain X is translated to domain Y and then back to domain X, it should ideally be identical to the original image. This constraint helps to preserve the underlying content and structure of the images. This consists of Forward Cycle and Backward Cycle which is mathematically represented as follows.

Forward Cycle: This term measures the L1 distance between the original image X and the reconstructed image $F(G(X))$. The L1 distance is used for simplicity, but other distance metrics can also be employed..$_1$ Represents the L1 distance (also known as the Manhattan distance or taxicab distance), which is the sum of the absolute differences between the corresponding pixel values of the original and reconstructed images.

$$L_{cyc_X} = ||X - F(G(X))||_1 \tag{2.15}$$

Backward Cycle: This term measures the L1 distance between the original image Y and the reconstructed image $G(F(Y))$.

$$L_{cyc_Y} = ||Y - G(F(Y))||_1 \tag{2.16}$$

The total CCL is the sum of the forward and backward cycle losses which is mathematically represented as shown in Eq. (2.17)

$$L_{cyc} = L_{cyc_X} + L_{cyc_Y} \tag{2.17}$$

The CCL plays a vital role in stabilizing the training process and ensuring the generated images maintain semantic coherence. By minimizing this loss, the generators are encouraged to learn mappings that are as close to identity functions as possible, preserving the underlying content of the images.

A well-balanced CCL is essential for generating high-quality and meaningful image translations.

2.3.1.3 Identity Loss

The identity loss in CycleGAN serves as a regularizer to encourage the generators to act as near-identity mappings when fed inputs from their respective domains [10]. This helps to preserve the original image content and prevent the generators from learning trivial mappings. Mathematically it can be represented as follows.

Identity Loss for Generator G: This term measures the L1 distance between the original image X from domain X and the image obtained by passing X through generator G and then back to domain X.

$$L_{id_x} = ||G(X) - X||_1 \tag{2.18}$$

Identity Loss for Generator F: This term measures the L1 distance between the original image Y from domain Y and the image obtained by passing Y through generator F and then back to domain Y.

$$L_{id_y} = ||F(Y) - Y||_1 \tag{2.19}$$

The identity loss helps to: (1) Preserve image content: By penalizing deviations from the original image, it encourages the generators to maintain the underlying structure and details, (2) Stabilize training: It can help prevent mode collapse and improve the overall performance of the CycleGAN model and (3) Enhance image quality: By ensuring that the generators can reconstruct the original images with high fidelity, it contributes to better image quality in the generated outputs.

2.3.1.4 Total Loss

The total loss for a CycleGAN is an amalgamation of adversarial loss, cycle consistency loss, and identity loss. The total loss for generator G is a weighted sum of the above losses and can be represented by using Eq. (2.20).

$$L_G = L_{G_X} + \lambda \times L_{cyc_X} + \lambda \times L_{id_X} \tag{2.20}$$

Similarly, the total loss for generator F is:

$$L_F = L_{G_Y} + \lambda \times L_{cyc_Y} + \lambda \times L_{id_Y}, \tag{2.21}$$

where λ is a hyperparameter that balances the contributions of the different loss terms.

The total loss for the discriminator is the sum of the losses for both discriminators and can be represented by using Eq. (2.22).

$$L_D = L_{D_X} + L_{D_Y} \qquad (2.22)$$

By minimizing these loss functions, the generators and discriminators are trained to produce high-quality image translations.

2.3.2 Training Process

CycleGAN training involves an iterative process of updating the generators and discriminators [11]. The goal is to minimize the overall loss function, which comprises adversarial, cycle consistency, and identity loss terms. The following steps are involved in the training process.

Data Preparation: First paired or unpaired image datasets are collected from two domains (X and Y). Then the preprocessing of the image is done such as resizing, normalization, etc.

Model Initialization: The generators G and F and discriminators Dx and Dy are initialized with random weights.

Training Loop: Randomly sample a batch of images from domains X and Y is considered for the training. In the forward pass the images are passed from domain X through generator G to obtain fake images in domain Y as represented in Eq. (2.23).

$$Y_{\text{fake}} = G(X) \qquad (2.23)$$

Then the images are passed from domain Y through generator F to obtain fake images in domain X as represented in Eq. (2.24).

$$X_{\text{fake}} = F(Y) \qquad (2.24)$$

After that, the real and fake images are passed from domain X to discriminator Dx and domain Y to discriminator Dy.

Next, the adversarial loss, cycle consistency loss, and identity loss (optional) are calculated for generators G and F. Then the total loss is computed for generators G and F. At last, the loss for discriminators Dx and Dy is computed. Depending on the loss the weights of generators G and F and discriminators Dx and Dy is calculated using backpropagation to minimize their respective total losses.

Hyperparameter tuning is essential for achieving optimal results [12]. The choice of generator and discriminator architectures can significantly impact performance. A diverse and representative dataset is essential for training a robust CycleGAN.

Figure 2.6 visualizes some images generated by CycleGAN after getting trained on the MNIST dataset.

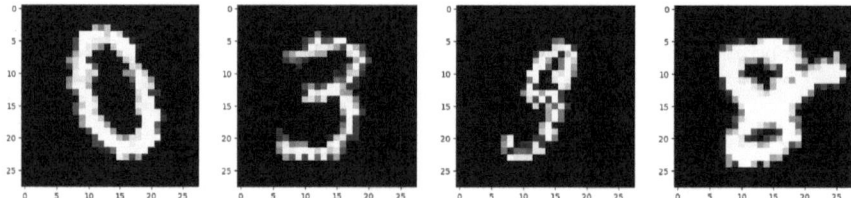

Fig. 2.6 Some sample images generated by CycleGAN

2.4 Super-Resolution GAN

Super-Resolution GAN (SRGAN) is a variant of GANs specifically designed for the task of image super-resolution [13]. It aims to generate high-resolution images from their low-resolution counterparts. The network consists of a generator and a discriminator. The architecture consists of one generator and one discriminator.

The generator in SRGAN is responsible for upscaling the low-resolution image to a high-resolution one. It typically consists of several convolutional and deconvolutional layers, along with nonlinear activation functions and normalization layers. A common architecture for the SRGAN generator includes the following components: (1) Shallow Feature Extraction: The initial convolutional layers of the SRGAN generator are responsible for extracting low-level features from the input low-resolution image. These features typically capture basic image characteristics such as textures, edges, and color gradients. This procedure is akin to the early stages of human visual perception, where we initially identify fundamental elements before processing more complex details. These extracted shallow features serve as the foundation for subsequent layers in the generator, which gradually build upon this information to construct the high-resolution output image. By effectively capturing low-level details, the generator can better reconstruct the missing information in the upscaling process. (2) Residual Blocks: A series of residual blocks are employed to capture deeper features and improve the learning process. Each residual block typically consists of two convolutional layers with batch normalization and ReLU activation. The output of the residual block is added to the input, forming a skip connection. This skip connection allows the network to learn residual mappings, which are the differences between the input and desired output. (3) Upsampling: Upsampling is a critical component of the SRGAN generator, responsible for increasing the spatial dimensions of the feature maps to match the desired high-resolution output. Two primary techniques are commonly employed sub-pixel convolution or transposed convolution can be used. Sub-pixel convolution, also known as pixel shuffle, is a computationally efficient method for upsampling. It involves rearranging the channels of a feature map into a higher-dimensional tensor. For instance, a feature map with shape $C \times H \times W$ can be reshaped into a feature map with shape $\left(C//r^2\right) \times rH \times rW$, where r is the upscaling factor. This reshaped feature map is then rearranged to form the final output with dimensions $C \times rH \times rW$. Sub-pixel convolution is effective in preserving fine details in the generated image. Transposed convolution, also known as deconvolution, is another technique for upsampling. It performs a convolution-like operation with learned weights to increase the spatial dimensions of the input feature map. While computationally more expensive than sub-pixel convolution, it can potentially learn more complex spatial relationships between pixels. (4) Reconstruction: The final stage of the SRGAN generator involves reconstructing the high-resolution image from the upsampled feature maps. This is typically achieved through a series of convolutional layers. The upsampled feature maps are fed into the first convolutional layer. Subsequent convolutional layers gradually refine the features and increase the spatial resolution of the output. The final convolutional layer produces the reconstructed high-resolution image with the same dimensions as the ground truth image.

The discriminator in SRGAN is responsible for distinguishing between real high-resolution images and those generated by the generator. It typically employs a convolutional architecture with pooling layers to downsample the input image. The discriminator receives a high-resolution image generated by the generator. A series of convolutional layers with increasing filter sizes and feature maps are applied to extract hierarchical features. LeakyReLU Activation is Introduced to introduce nonlinearity and prevent vanishing gradients. Strided convolutions or pooling layers are utilized to decrease the spatial dimensions of the feature maps. Lastly, flatten the feature maps and pass them through fully connected layers to produce a final representation. The output layer is a single sigmoid activation unit that outputs the probability of the input image being real. PatchGAN Discriminator is often used in SRGAN, the discriminator is applied to patches of the image rather than the entire image.

The discriminator's loss function is a binary cross-entropy loss. The discriminator aims to maximize this loss function, while the generator aims to minimize it.

2.4.1 Loss Function

The overall loss function for SRGAN consists of three components [14]: (1) Adversarial Loss: Encourages the generator to produce images indistinguishable from real high-resolution images and uses a binary cross-entropy loss for the discriminator. The generator aims to minimize the discriminator's accuracy. (2) Content Loss: Measures the pixel-wise difference between the generated image and the ground truth high-resolution image and uses L1 or L2 loss. (3) Perceptual Loss: Encourages the generator to produce images with perceptual features similar to the ground truth, Uses a pre-trained image classification network (e.g., VGG) to extract feature maps, and Calculates the L1 or L2 distance between the feature maps of the generated and ground truth images. The total loss function for the generator is a weighted sum of these three losses.

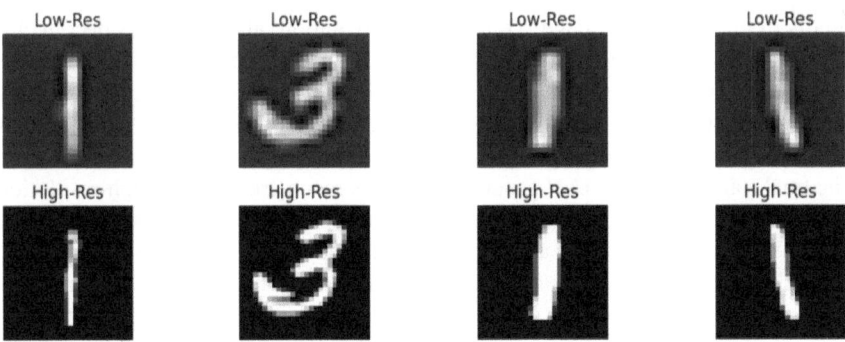

Fig. 2.7 The high-resolution (high-res) images generated from low-resolution (low-res) images by the SRGAN

2.4.2 Training Process

The SRGAN training process involves an iterative adversarial learning approach. Initially, the generator and discriminator are randomly initialized [15]. In each iteration, a batch of low-resolution and corresponding high-resolution image pairs is fed into the network. The generator upscales the low-resolution image and the discriminator classifies the generated image as real or fake. Based on the discriminator's output, both the generator and discriminator are updated using backpropagation to minimize their respective loss functions. The generator aims to produce images that can fool the discriminator, while the discriminator strives to accurately differentiate between real and fake images. This adversarial process continues until the generator produces high-quality super-resolved images that are indistinguishable from real images. Figure 2.7 displays some high-resolution (high-res) images generated from low-resolution (low-res) images by the SRGAN. I trained the SRGAN on the MNIST dataset.

2.5 Applications of GAN in Image Augmentation

GANs have found applications across various domains, using their capability to produce images based on definite conditions.

2.5.1 Image-To-Image Translation

Image-to-image translation is a powerful application of cGANs [16]. It involves transforming an input image from one domain to another while preserving semantic information. This can be anything from converting photos to paintings, translating day scenes to night scenes, or generating maps from aerial images. By conditioning the generator and discriminator on both the input image and the desired output domain, cGANs learn to map complex image structures from one domain to another. Key techniques like paired image datasets, image-specific losses, and architectural considerations such as U-Net-based generators and PatchGAN discriminators contribute to the success of these models.

2.5.1.1 Style Transfer

Style transfer aims to transfer the artistic style of one image (style image) onto the content of another image (content image) [17, 18]. This process involves combining the low-level features (style) of one image with the high-level features (content) of another. To capture the image content for style transfer, a pre-trained convolutional neural network (CNN), such as VGG, is employed. This network, trained on a massive dataset for image classification, has learned to extract hierarchical features from images. By feeding the content image through this network, we obtain feature maps at different layers, each representing different levels of abstraction. To encapsulate the content information of the image, the activations of a specific intermediate layer, often referred to as the content layer (e.g., conv4_2 in VGG), are extracted. These activations serve as a numerical representation of the content, capturing the essential structural information of the image while discarding unnecessary details. Essentially, the content image is transformed into a feature space where it is represented by a set of feature maps, and these feature maps are the core of the content representation used in style transfer.

To capture the style of an image, the Gram matrix is employed. This matrix measures the correlation between different feature channels within a specific layer of a convolutional neural network. By calculating Gram matrices for multiple layers, we can capture a comprehensive representation of the image's style, from low-level textures to high-level composition. The Gram matrix of a feature map is computed as the outer product of the feature map with itself, followed by a normalization step. Mathematically, if F is a feature map of size $C \times H \times W$ (channels, height, width), the Gram matrix G is calculated as shown in Eq. (2.25).

$$G_{ij} = \text{sum}_{\{k,l\}} F_{\{ik,jl\}}, \tag{2.25}$$

where: G_{ij} is the element at position (i, j) of the Gram matrix and $F_{\{ik,jl\}}$ is the element at position (i, k) and (j, l) of the feature map.

By computing Gram matrices for different layers of the CNN applied to the style image, we obtain a set of style representations that encapsulate the image's artistic characteristics. These style representations will be used to guide the generation of the output image, ensuring it adopts the desired style.

The generator in a style transfer network is tasked with the challenging problem of merging the content of one image with the style of another. It receives as input the content features extracted from the content image and the style features encoded in the Gram matrices. The generator's architecture, which is based on convolutional neural networks, processes this information and produces an output image that ideally preserves the semantic and structural content of the original image while adopting the artistic style of the style image. This involves a complex interplay between the low-level features learned from the content image and the high-level style patterns captured in the Gram matrices. The generator must learn to balance these two components to produce a visually pleasing and coherent output image.

The content loss in style transfer aims to preserve the structural and semantic information of the original content image within the generated output. This is achieved by measuring the difference between the content features extracted from the original image and those extracted from the generated image as shown in Eq. (2.26).

$$L_{\text{content}} = \Phi_c(x) - \Phi_c(G(x_c, s))_2, \tag{2.26}$$

where: Φ_c is the content feature extractor, x is the content image, $G(x_c, s)$ is the generated image, s is the style image, $||.||_2$ is the L2 norm.

The style loss is a critical component of neural style transfer, quantifying the dissimilarity between the style of the generated image and the desired style image. Unlike content loss, which focuses on structural similarity, style loss captures the artistic essence, including color palette, texture, and brushstroke-like patterns. To compute style loss, Gram matrices are employed as shown in Eq. (2.27).

$$L_{\text{style}} = \Sigma_l ||G_l(x_c, s) - S_l||_F^2 \tag{2.27}$$

where: $G_l(x_c, s)$ is the Gram matrix of the $l - $ th layer of the generated image, S_l is the Gram matrix of the $l - $ th layer of the style image, $||.||_F^2$ is the Frobenius norm.

By combining content loss and style loss, the generator learns to create images that harmoniously blend the structure of one image with the artistic flair of another as shown in Eq. (2.28).

$$L_{\text{total}} = \alpha L_{\text{content}} + \beta L_{\text{style}}, \tag{2.28}$$

where: α and β are weights for content and style loss respectively.

While the basic style transfer process involves transferring the style of one image to the content of another, several advanced techniques have been explored to improve the quality and versatility of the generated images. Here are a few examples of this.

Deep Feature Extraction: While Gram matrices effectively capture low-level style characteristics, deeper features extracted from convolutional neural networks offer a richer representation of the image's style. By analyzing features from multiple layers of the network, we can capture a more comprehensive understanding of the artistic essence. This approach, often referred to as deep feature extraction, allows for a more nuanced style transfer, preserving not only the superficial style elements but also the underlying artistic intent. By computing similarity metrics between the deep features of the generated image and the style image, the generator can be trained to reproduce the target style with greater fidelity, resulting in more visually appealing and authentic style-transferred images.

Attention Mechanisms: Attention mechanisms have significantly enhanced the capabilities of style transfer models by enabling a more focused and nuanced transfer of style information [19]. By directing the generator's attention to specific regions of the style image, these mechanisms allow for a more precise and detailed style transfer. Attention mechanisms generate attention maps that highlight the most relevant regions in the style image for transferring specific style elements. The generator uses these attention maps to weight the contribution of different style features to the generated image. By focusing on specific regions, attention mechanisms help to preserve fine details and prevent style distortions. In this way more accurate and precise transfer of style information is possible. Also, artifacts and distortions in the generated images can be reduced. This technique can be adapted to various style transfer tasks, including those with complex style patterns.

Style-Specific Normalization: Instance normalization is a technique that normalizes the activations of a layer for each image, rather than across a batch of images as in batch normalization [20]. This is particularly useful in style transfer as it helps to preserve style-specific information within an image. By applying instance normalization to the generator's layers, we can enhance the transfer of stylistic elements, such as color palette, texture, and brushstrokes, from the style image to the generated output. This technique helps to maintain the distinctiveness of the style while preserving the content information of the original image. By normalizing the activations at the instance level, the generator can focus on learning the style-specific patterns without being influenced by the statistics of other images in the batch. This leads to a more accurate and effective style transfer process.

Hierarchical Style Transfer: Hierarchical style transfer involves applying style transfer at multiple scales to capture both fine-grained and coarse-grained style details [21]. By decomposing the style image into different levels of abstraction, the generator can learn to transfer specific style elements at each level. This approach enhances the quality of the generated image by preserving both the overall artistic impression and the intricate details of the style. By processing the style image through a convolutional neural network, feature maps at different layers can be extracted, representing different levels of style information. These feature maps are then used to guide the style transfer process at corresponding scales in the generator. This hierarchical method allows for a more accurate and nuanced transfer of style, resulting in visually appealing and high-fidelity output images.

Pyramid Style Transfer: Pyramid style transfer is a technique that involves decomposing the style image into multiple scales, each representing different levels of detail [22]. By applying style transfer independently at each scale, the generator can capture a more comprehensive and nuanced representation of the style. This approach is particularly effective in preserving fine-grained details while maintaining the overall artistic coherence of the generated image. By constructing a pyramid of style features from the style image, the generator can progressively transfer style information from coarse to fine levels, resulting in a more accurate and visually pleasing output. This hierarchical approach allows for more granular control over the style transfer procedure, enabling the creation of highly detailed and aesthetically pleasing images.

Content Loss Refinement with Perceptual Loss: While traditional content loss focuses on pixel-level differences, perceptual loss takes a more holistic approach by comparing high-level features extracted from a pre-trained convolutional neural network (CNN) [23]. This method aligns more closely with human perception, as we often perceive images based on their overall structure and content rather than individual pixel values. By computing the distance between feature maps of corresponding layers in the generated and target images, perceptual loss encourages the generator to preserve not just the low-level details but also the semantic and structural information of the content image. This leads to generating images that are perceptually more similar to the original, resulting in a more natural and visually pleasing output.

Semantic Alignment in Style Transfer: Preserving the semantic meaning of the content image during style transfer is crucial for generating visually pleasing and meaningful results [24]. Semantic alignment ensures that the objects and their spatial relationships in the content image are maintained in the stylized output. Several techniques have been explored to achieve semantic alignment. By aligning the feature representations of the content and style images, it's possible to preserve semantic information. Methods like correlation-based alignment or subspace learning can be employed to find corresponding features in both images. Using semantic segmentation to identify different objects in the content image and then applying style transfer to each object separately can help preserve object boundaries and relationships. Incorporating adversarial losses that penalize changes in object recognition accuracy can encourage the generator to maintain semantic consistency. By focusing on semantic alignment, style transfer models can produce more realistic and visually appealing results, avoiding distortions that might compromise the underlying meaning of the content image.

Style transfer, powered by GANs, has opened up a world of creative possibilities. Here are some specific applications.

Artistic Transformations: GANs have revolutionized the field of image stylization, enabling the transformation of photographs into artistic masterpieces reminiscent of renowned painters like Monet, Van Gogh, or Picasso [25]. By training a GAN on a dataset of images paired with their corresponding stylized versions, the model learns to capture the essence of a particular artistic style. When presented with a new photograph, the generator component of the GAN can effectively apply the learned style, transforming the ordinary into the extraordinary. This process involves extracting the content from the input photograph while overlaying the stylistic characteristics of the chosen artist, resulting in a visually striking and aesthetically pleasing output. The real-time application of this is as follows. Imagine an application that allows users to instantly transform their photos into works of art reminiscent of famous painters. By using GAN technology, users could select from a variety of artistic styles, such as Van Gogh, Monet, or Picasso, and apply them to their images in real-time. This app could offer features like style blending, intensity control, and even custom style creation. Such a tool would democratize art, allowing anyone to experience the joy of artistic expression and creation. This application could be integrated into social media platforms, enabling users to share their artistic creations and engage with a community of art enthusiasts. Additionally, it could be used by artists and designers as a source of inspiration and a tool for rapid prototyping.

Design Pattern Generation: GANs have shown remarkable potential in creating innovative design patterns for textiles, wallpapers, and graphic design [26]. By training a GAN on a diverse dataset of existing patterns, the model learns to capture the underlying style, structure, and complexity of these designs. The generator component of the GAN then produces new patterns by manipulating random noise inputs, aiming to mimic the real patterns. This adversarial process between the generator and discriminator leads to the creation of highly realistic and visually appealing design patterns. Techniques like conditional GANs can be employed to control the generated patterns based on specific style preferences or design constraints. However, ensuring seamless pattern repetition, maintaining aesthetic appeal, and addressing potential copyright issues are crucial considerations in this domain. The real-time application of this is as follows. Imagine an application that allows designers to instantly generate a variety of design patterns based on a chosen style or theme. Using GAN technology, the app could offer a vast library of pattern styles, ranging from geometric to floral, abstract to ethnic. Designers could input specific color palettes, pattern scales, or even sketches as prompts, and the app would generate multiple design variations in real-time. This tool would be invaluable for fashion designers, interior decorators, and graphic artists, accelerating the design process and inspiring creativity. Additionally, the app could offer features like pattern customization, saving, and sharing, enhancing the user experience.

Logo Generation: GANs have demonstrated their potential in generating logo variations based on different artistic styles [27]. By training a GAN on a dataset of logos and their corresponding stylistic variations, the model learns to capture the essence of various logo styles. The generator can then produce new logo designs by combining the core elements of a base logo with different artistic styles. This approach offers a creative and efficient way to explore diverse logo options, permitting designers to experiment with different visual aesthetics without manual intervention. However, ensuring consistency with brand identity and maintaining logo readability remain crucial challenges in this domain. The real-time application of this is as follows. Imagine a design tool that allows users to input a basic logo concept or even a hand-drawn sketch, and then generates a variety of logo options in different styles. This is where GANs excel. By training a model on a vast dataset of logos across various industries and styles, a generator can be taught to produce new logo designs based on user-specified parameters. For instance, a user might input a simple geometric shape as a base logo and select a desired style (e.g., minimalist, retro, or futuristic). The GAN would then generate multiple logo variations incorporating the chosen style while maintaining the core elements of the input shape. This empowers designers to rapidly explore different visual directions and find the perfect logo for their brand.

Image Enhancement: GANs have demonstrated remarkable capabilities in enhancing image quality by applying artistic filters or boosting specific features

[28]. By training a GAN on pairs of low-quality and high-quality images, the generator learns to map low-quality inputs to their enhanced counterparts. This involves capturing the underlying details and structure of the low-quality image while introducing improvements such as increased sharpness, reduced noise, or enhanced color saturation. The discriminator acts as a critic, evaluating the quality of the generated enhanced images and providing feedback to the generator. Through this adversarial process, the GAN can produce visually appealing and high-quality enhanced images. Techniques like perceptual loss, which focuses on preserving high-level image features, can be incorporated to maintain the overall image content while improving its appearance. Additionally, conditional GANs can be used to control the enhancement process based on specific requirements, such as increasing contrast or reducing noise. By using the power of GANs, it's possible to achieve significant improvements in image quality while preserving the original image content. The real-time application of this is as follows. Imagine an application capable of instantly enhancing low-quality or poorly lit photos. Powered by a GAN specifically trained for image enhancement, this app could significantly improve image quality, reducing noise, increasing sharpness, and restoring details. Users could apply various enhancement levels, from subtle adjustments to dramatic transformations. Such an app would be invaluable for photographers, social media users, and anyone looking to improve their image quality on the go. Additionally, the app could offer advanced features like selective enhancement, allowing users to focus improvements on specific image regions. This level of control would provide users with greater flexibility and customization options.

Image Restoration: GANs have shown promise in restoring damaged or degraded images by effectively transferring information from similar, undamaged images [29]. This process involves training a GAN on pairs of degraded and corresponding restored images. The generator learns to map degraded images to their restored counterparts, capturing underlying patterns and structures. By incorporating style transfer techniques, the GAN can further enhance the restoration process by transferring visual styles from undamaged images to the restored output. This combined approach allows for the recovery of lost details, reduction of noise, and improvement of overall image quality while preserving the original image content. The real-time application of this is as follows. Imagine an application designed to restore old, damaged, or low-resolution photographs. By using the power of GANs, this app could effectively remove noise, scratches, and other imperfections, bringing old memories back to life. Users could upload their cherished photos, and the app would apply advanced restoration techniques to enhance image quality, recover lost details, and restore colors. This technology could be particularly beneficial for preserving historical photographs and family heirlooms. Additionally, the app could offer features like upscaling to improve image resolution and selective restoration, allowing users to focus on specific areas of the image. Such an app could revolutionize the way we preserve and enjoy our photographic memories, making it possible to restore even severely damaged images to a remarkable degree of clarity and detail.

Colorization: GANs have proven effective in the task of image colorization, transforming black and white images into vibrant color versions [30]. By training a cGAN on a dataset of grayscale and corresponding color images, the model learns to infer plausible colors based on the image content. The generator component of the GAN takes a grayscale image as input and produces a colorized output. The discriminator differentiates between real color images and the generated colorized images, driving the generator to produce increasingly accurate and visually pleasing results. This approach has shown promising results in capturing the essence of the original scene while adding realistic colors. While the core concept is relatively straightforward, challenges such as preserving color harmony, handling shadows and highlights, and ensuring color consistency require careful consideration in the GAN architecture and training process. The real-time application of this is as follows. Imagine an application that can instantly transform black and white photographs into vibrant color images. Using the power of GANs, this app could analyze the grayscale image and intelligently apply colors based on the content and context. Users could experiment with different color palettes or styles, creating a wide range of colorized versions of their black-and-white photos. This technology would be particularly useful for restoring old photographs, enhancing artistic expression, and creating visually appealing content for social media. The app could also incorporate features like selective colorization, allowing users to colorize specific parts of an image while leaving others in black and white. Additionally, advanced techniques like color harmony and consistency could be implemented to ensure visually pleasing results.

2.5.1.2 Photo-to-Cartoon

GANs have shown remarkable capabilities in transforming real-world photographs into stylized cartoon-like images [31]. By training a GAN on pairs of real images and their corresponding cartoon counterparts, the model learns to capture the essence of cartoonization. The generator in the GAN is tasked with converting a given photograph into a visually appealing cartoon, while the discriminator differentiates between real cartoon images and those generated by the generator. This adversarial process refines the generator's ability to produce high-quality cartoon images that preserve key features of the original photograph while introducing a distinct cartoonish style. Several GAN architectures, such as CycleGAN and Pix2Pix, have been successfully applied to this task, demonstrating impressive results in terms of style preservation and image quality.

Converting a photograph into a cartoon involves a complex transformation that requires preserving essential features while introducing a stylized appearance. This transformation involves changes in the color palette, edge enhancement, and simplification of details.

One real-time example of this is as follows. Imagine a mobile application that permits users to instantly transform their photos into captivating cartoons. Powered by a GAN model trained on a vast dataset of images and their corresponding cartoon counterparts, this app could offer a range of artistic styles, from classic anime to modern comic book aesthetics. Users could experiment with different styles, adjust parameters like line thickness and color saturation, and share their creations on social media. Such an app would not only be a fun and engaging tool for casual users but also a valuable asset for artists and designers seeking inspiration or quick concept visualizations.

The following are the challenges and considerations of Photo-to-Cartoon using GAN [32].

Preservation of Key Features: Capturing and preserving essential facial features, object shapes, and overall composition in the cartoonized output is crucial. Loss functions that focus on structural similarity can help address this.

Stylization: Achieving a convincing cartoon style, including simplified shapes, bold outlines, and appropriate color palettes, is essential. Techniques like edge enhancement and color quantization can be employed.

Dataset Bias: The quality of the generated cartoons is influenced by the diversity and quality of the training dataset. A balanced dataset with various cartoon styles is necessary to prevent overfitting.

Computational Resources: Training GANs for photo-to-cartoon conversion can be computationally intensive, necessitating significant hardware resources.

Evaluation Metrics: Assessing the quality of cartoonized images is challenging due to the subjective nature of artistic style. Perceptual metrics and user studies can be used to evaluate the results.

To elevate the quality of photo-to-cartoon conversions using GANs, a multifaceted approach is essential. Techniques such as perceptual loss, which focuses on higher-level image features, can be integrated to preserve semantic and structural information beyond pixel-level details. Additionally, integrating attention mechanisms can direct the generator to emphasize specific regions, refining the transfer of stylistic elements. To further enhance the cartoonish appearance, edge preservation techniques can be employed to accentuate outlines, while color quantization can reduce the color palette for a more pronounced cartoon effect. By combining these strategies with careful dataset curation and hyperparameter tuning, it's possible to achieve highly realistic and visually appealing cartoonized images.

2.5.1.3 Domain Transfer of an Image

Domain transfer of an image involves transforming an image from one domain (e.g., day scene) to another (e.g., night scene) while preserving semantic information. This challenging task has been revolutionized by GANs [33]. CycleGAN is particularly suited for this, as it doesn't require paired training data. Here adversarial loss encourages realistic image generation, while cycle consistency ensures semantic preservation.

Here are some real-time applications of this: (1) Real-time Weather App: Imagine a weather app that not only provides weather forecasts but also visually transforms real-time camera feeds to simulate different weather conditions. Using GAN-based image-to-image translation, the app could convert a sunny day scene into a rainy, snowy, or cloudy scenario in real-time. This would provide users with a more immersive weather experience, helping them visualize the potential impact of different weather conditions on their surroundings. For instance, a user could see how a sunny park would look under heavy rain or snowfall, aiding in planning outdoor activities. (2) Real-time Product Visualization: Imagine a mobile app for e-commerce platforms that allows users to visualize products in different environments or with varying lighting conditions. Using image-to-image translation powered by GANs, the app could transform a product image from a studio setting to a living room or outdoor environment. This would provide customers with a more realistic and immersive shopping experience, helping them envision how the product would look in their own space. For instance, a user could see how a particular sofa would look in their living room before making a purchase.

Image-to-image translation with GANs faces several challenges [34]. The generator may converge to produce a limited set of similar images, reducing diversity. Ensuring that the translated image accurately reflects the semantic content of the original image is crucial. Generating highly realistic and visually pleasing images in the target domain is often challenging. Training GANs for image-to-image translation can be computationally expensive because of the large number of parameters and iterations. Sufficient and diverse paired or unpaired training data is essential for achieving good results. Developing reliable metrics to assess the quality of image-to-image translation remains an open challenge.

Overcoming challenges in image-to-image translation requires a multifaceted approach [35]. Methods like spectral normalization, instance normalization, and attention mechanisms can support stabilizing training and enhance image quality. Incorporating auxiliary classifiers or semantic segmentation information can enhance semantic preservation. To mitigate mode collapse, diverse datasets, careful hyperparameter tuning, and the use of regularization techniques are crucial. Additionally, exploring alternative loss functions, like feature matching or perceptual loss, can lead to more meaningful image translations. By combining these strategies, researchers aim to develop robust and versatile image-to-image translation models.

2.5.2 Image Generation

GANs have revolutionized image generation. GANs have found applications in various fields. Some of them are as follows.

2.5.2.1 Conditional Image Generation

GANs extend the capabilities of traditional GANs by incorporating additional information, known as conditioning information, to guide the image-generation process [36]. This is particularly done by cGans. The conditioning information can be in the form of class labels, attributes, textual descriptions, or other relevant data. By providing this extra context, cGANs can generate images that adhere to specific conditions, offering greater control and flexibility compared to unconditional GANs. One of the most significant applications for generating images from class labels is data augmentation. By creating synthetic images with specific class labels, we can expand the diversity and size of training datasets. This is mainly useful for datasets with imbalanced class distributions or limited data availability. For instance, in medical image analysis, generating synthetic images of rare diseases can help improve model performance. By employing cGANs, textual dream descriptions can be translated into visual representations. The dream narrative serves as the conditioning information, guiding the generator to produce images that align with the dream's content. This process involves encoding the textual description into a suitable format, such as word embeddings, and feeding it as input to the generator along with random noise. The generator then produces an image that embodies the essence of the dream, offering a unique and personalized interpretation of the dreamer's subconscious thoughts. Generating product designs based on textual descriptions or category labels is a promising application of conditional GANs. By providing a textual description of a product, such as "a modern, minimalist chair," the cGAN can generate multiple design variations. This approach accelerates the design process, allowing designers to explore a wide range of possibilities efficiently. Additionally, by incorporating user preferences or style guidelines into the textual description, the generated designs can be tailored to specific requirements. This technology has the potential to revolutionize product design by providing designers with a powerful tool for ideation and exploration. GANs have shown immense potential in accelerating game development by automating the creation of diverse game assets. By training a GAN on a dataset of existing game assets, such as characters, environments, or items, the model can learn to generate new, realistic, and varied assets. This can significantly decrease the effort and time required by artists and designers. For instance, a GAN can generate multiple character variations based on a given character class, or create diverse environments for different game levels. This approach not only enhances productivity but also fosters creativity by providing a vast pool of assets for experimentation. However, ensuring consistency in style, gameplay relevance, and optimization for game engines remain crucial challenges in this domain. By using the power of cGANs, we can generate a vast array of images based on textual or categorical information, opening up new possibilities for creativity and problem-solving.

Even though there are various applications of conditional image generation using GAN, it faces some challenges also. The generator might converge to produce only a limited set of images for a specific condition, reducing diversity. If the training data is imbalanced across different classes, the generator might struggle to generate high-quality images for underrepresented classes. Effectively encoding and utilizing complex conditioning information can be challenging, especially for multimodal or hierarchical conditions. Assessing the superiority of generated images based on conditions requires specialized metrics and human evaluation. Training cGANs can be computationally expensive, especially for large datasets and complex models.

Overcoming challenges in conditional image generation requires a multifaceted approach. To address mode collapse, techniques like spectral normalization, label smoothing, and careful hyperparameter tuning are essential. For imbalanced datasets, data augmentation, oversampling, and class weighting can be employed. Effective encoding of complex conditioning information can be attained through methods like one-hot encoding, embedding layers, or attention mechanisms. To evaluate generated images accurately, a combination of quantitative metrics (e.g., Inception Score, Fréchet Inception Distance) and human evaluation is recommended. By merging these approaches, we can develop robust and effective conditional image-generation models.

2.5.2.2 Super-Resolution

Super-resolution (SR) is the procedure of improving the resolution of a low-resolution image to produce a higher-resolution image. Traditional methods often resulted in blurry or pixelated outputs. However, the advent of SRGAN has revolutionized this field.

One of the most promising applications of SRGAN is in real-time video enhancement. By applying the super-resolution technique to each frame of a video, it's possible to significantly improve the overall video quality. This has implications for various industries. GANs offer a promising solution for enhancing video conferencing quality by addressing the limitations of traditional compression techniques [37]. By applying super-resolution techniques based on GANs, it's possible to upscale the resolution of each video frame in real-time, resulting in a significantly improved visual experience. This includes training a GAN model on a dataset of low-resolution and corresponding high-resolution video frames. The generator learns to produce high-resolution outputs from low-resolution inputs, while the discriminator distinguishes between real and generated frames. By iteratively improving the generator's ability to create realistic high-resolution images, GANs can effectively enhance video call clarity, reducing pixelation and improving overall image quality, leading to a more immersive and engaging video conferencing experience. Surveillance cameras often capture low-resolution footage, hindering effective analysis and identification. GANs, particularly SRGAN, offer a promising solution to this challenge. By training a GAN on pairs of low-resolution and high-resolution images, the model learns to generate detailed and sharp images from low-quality input. This enhancement significantly improves the ability to recognize individuals, vehicles, and objects in surveillance footage, aiding in crime prevention, investigation, and public safety. Additionally, GANs can be used to address other surveillance challenges, such as low-light conditions and occlusion, by generating enhanced images under various conditions. By applying SRGAN techniques, video streaming platforms can significantly improve viewer experience by upscaling low-resolution content to higher resolutions in real-time. This is particularly beneficial for older content or videos originally produced in standard definition. The upscaled videos appear sharper, with enhanced details and reduced pixelation, providing a more immersive viewing experience, especially on larger screens and high-definition displays. This technology can breathe new life into video archives and expand the appeal of older content to a wider audience. Additionally, for users with limited bandwidth, super-resolution can be utilized to improve the perceived quality of lower-resolution streams, making them more enjoyable to watch.

2.5.2.3 Image Inpainting

Image inpainting is the process of filling in missing parts of an image. GANs have demonstrated remarkable capabilities in this domain [38]. By training a GAN on a dataset of images with masked regions, the generator learns to reconstruct the missing parts while preserving the overall image consistency. The generator takes an image with missing regions as input and generates a complete image. The discriminator distinguishes between real images and images with inpainted regions. Loss function typically includes adversarial loss, content loss, and perceptual loss to guide the training process.

One real-time application of this is as follows. Imagine a mobile application capable of restoring damaged photos in real-time. Using GAN-based image inpainting, users could remove scratches, tears, or water damage from their old photographs. The app could also be used to restore faded colors and enhance image details. This would be invaluable for preserving family heirlooms and historical photographs. Additionally, the app could offer advanced features like object removal, where users can seamlessly erase unwanted elements from their images.

Image inpainting using GANs faces several challenges [38, 39]. Some of them are as follows.

Preserving Image Coherence: Maintaining consistency between the inpainted region and the original image is a critical challenge in image inpainting. The generated content must seamlessly blend with the existing image, avoiding noticeable artifacts or distortions. To achieve this, techniques such as attention mechanisms, which focus on relevant image regions, and partial convolution, which handles missing pixels effectively, can be employed. Additionally, incorporating style transfer techniques can help maintain the overall visual style of the image while filling in missing details. By carefully considering these factors, it is possible to generate inpainted images that are visually indistinguishable from the original content.

Handling Large Missing Regions in Image Inpainting: Inpainting large missing regions in images is a particularly challenging task for GANs. Accurately estimating missing details and structures while maintaining consistency with the surrounding image content requires sophisticated techniques. Traditional inpainting methods often struggle with such scenarios, leading to noticeable artifacts or unnatural completions. To address this, advanced GAN architectures and loss functions are necessary. Techniques like partial convolution, attention mechanisms, and hierarchical representations can help capture global and local context, enabling the generator to produce more plausible and coherent inpainted regions. Additionally, incorporating semantic information about the missing content can improve the accuracy of the inpainting process.

Generating Realistic Details in Image Inpainting: Creating plausible and visually appealing content for missing image regions is a critical challenge in image inpainting. The inpainted content must seamlessly blend with the existing image, maintaining consistency in terms of color, texture, lighting, and object structures. To achieve this, GANs must be trained to understand and replicate complex image patterns. Techniques like attention mechanisms can help focus the generator on relevant image regions, while perceptual loss can guide the generation of content that is perceptually similar to the original image. Additionally, incorporating style transfer concepts can enhance the aesthetic quality of the inpainted region. By carefully considering these factors, it's possible to produce highly realistic and visually convincing inpainted images.

Several techniques have been developed to improve the performance of GAN-based image inpainting. Some of them are as follows.

Partial Convolution: Partial convolution is a key technique for addressing the challenge of large missing regions in image inpainting [40]. Focusing computations on valid image pixels, prevents the propagation of information from missing areas, thereby reducing artifacts. This method involves creating a mask to identify valid pixels, performing convolutions only on these regions, and scaling the output based on the number of valid pixels. By incorporating partial convolution into GAN-based inpainting models, it is possible to achieve more accurate and realistic reconstructions, particularly when dealing with extensive image damage.

Contextual Attention: Contextual attention is a technique that uses information from surrounding regions to enhance image inpainting [41]. It addresses the limitation of convolutional neural networks in capturing long-range dependencies. By attending to relevant parts of the image, this method helps to generate more realistic and coherent inpainted regions. The core idea involves using features from known image patches as convolutional filters to process the generated patches. This process includes convolution for matching generated patches with known contextual patches, channel-wise softmax for weighting relevant patches, and deconvolution for reconstructing the generated patches with contextual information. Additionally, a spatial propagation layer encourages spatial coherence in the attention map. By incorporating contextual attention, the inpainting model can effectively borrow information from distant spatial locations, leading to improved generation of missing image content and a more natural appearance of the inpainted region.

Generative Adversarial Networks with Auxiliary Classifiers (ACGANs): ACGANs enhance image inpainting by incorporating class labels as auxiliary information [42]. The generator in an ACGAN receives both random noise and class labels as input, generating images conditioned on the specified class. The discriminator, in addition to determining image authenticity, also predicts the class label of the input image. This dual task improves the generator's ability to produce images that align with the given class, leading to more accurate and detailed inpainted regions. By incorporating class information, ACGANs can better understand the semantic content of the image, resulting in more coherent and realistic inpaintings. However, training ACGANs requires careful balancing of the adversarial and classification losses, and it's essential to handle imbalanced datasets effectively. Additionally, encoding complex class information might pose challenges in certain applications.

Multi-scale Approach: A multi-scale approach involves processing the image at multiple resolutions to capture both fine-grained and coarse-grained details [43]. In the context of GAN-based image inpainting, this technique can be implemented by feeding the image through a pyramid of image resolutions. At each level, a separate inpainting network can be applied, focusing on specific details. The outputs from different scales can then be fused to generate a final, high-quality inpainted image. By operating at multiple scales, the model can effectively address challenges such as preserving fine-grained textures while maintaining overall image consistency. Additionally, this approach can improve the handling of large missing regions by providing a hierarchical representation of the image content.

Generative Adversarial Networks with Conditional Instance Normalization (CIN): Conditional Instance Normalization (CIN) is a technique that enhances the performance of GANs, particularly in tasks like image inpainting [44]. By conditioning the instance normalization parameters on the input image, CIN allows the network to capture style-specific information and generate more diverse and realistic inpainted regions. Typically, instance normalization normalizes the activations of each channel within an image independently. CIN extends this by introducing learnable scale and shift parameters for each channel, conditioned on the input image. These parameters are learned during training to capture style-specific information. CIN is integrated into the generator network, allowing it to adapt the normalization process based on the input image content. By adapting the normalization parameters to the input image, CIN promotes the generation of diverse and realistic inpainted regions. CIN helps the generator handle variations in lighting, texture, and other image characteristics, leading to more robust inpainting results.

2.6 Challenges and Future Scopes of GAN-Based Image Augmentation

GANs have shown immense potential in image augmentation, but several challenges persist [45]. Primarily, GANs are notorious for instability during training, often leading to mode collapse where the generator generates a limited set of images. Additionally, ensuring the generated images accurately represent the underlying data distribution is crucial. Generating diverse and realistic augmentations while maintaining semantic consistency remains a challenge. Furthermore, assessing the quality of augmented images is subjective and lacks standardized metrics.

Despite challenges, GANs offer promising avenues for future research in image augmentation. Developing more stable GAN architectures and training techniques is a priority. Incorporating auxiliary information, such as semantic segmentation masks or object detection annotations, can enhance the realism and diversity of generated images. Exploring conditional GANs for targeted augmentation based on specific image attributes is another promising direction. Additionally, research on evaluating the effectiveness of augmented data on downstream tasks is essential to understand the true impact of GAN-based augmentation. By addressing these challenges and exploring new avenues, GANs can become even more powerful tools for augmenting image datasets and improving machine learning models.

2.7 Summary

In this chapter different aspects of GAN-based image augmentation are discussed. GANs have emerged as a powerful tool for image augmentation, surpassing traditional methods in generating realistic and diverse synthetic images. By pitting a generator against a discriminator, GANs learn to create images that closely resemble real data, expanding training datasets and enhancing model performance. Key applications include super-resolution, image-to-image translation (e.g., day-to-night, photo-to-cartoon), image inpainting, and generating images from class labels or textual descriptions. GANs have also shown promise in generating game assets and enhancing image quality through techniques like style transfer and colorization. Challenges in GAN-based image augmentation include mode collapse, preserving semantic information, generating realistic details, computational cost, and evaluation metric development. Addressing these challenges requires advanced techniques like attention mechanisms, partial convolution, and style transfer integration. Future directions involve developing more stable GAN architectures, incorporating auxiliary information, and exploring effective evaluation metrics. By overcoming current limitations, GANs have the potential to become indispensable tools for augmenting image datasets and improving machine learning models across various domains.

References

1. Goodfellow, I., et al. (2014). Generative adversarial nets. *Advances in neural information processing systems, 27.*
2. Wang, K., Gou, C., Duan, Y., Lin, Y., Zheng, X., & Wang, F. Y. (2017). Generative adversarial networks: Introduction and outlook. *IEEE/CAA Journal of Automatica Sinica, 4*(4), 588–598.
3. Huang, Z., et al. (2023). What can discriminator do? towards box-free ownership verification of generative adversarial networks. In *Proceedings of the IEEE/CVF international conference on computer vision* (pp. 5009–5019).
4. Jia, Q., & Ma, Z. (2020, March). Patch-based generative adversarial network for single image haze removal. In *2020 International conference on computer engineering and application (ICCEA)* (pp. 882–886). IEEE.
5. Thekumparampil, K. K., Khetan, A., Lin, Z., & Oh, S. (2018). Robustness of conditional gans to noisy labels. *Advances in neural information processing systems, 31.*
6. Abdelmotaal, H., Abdou, A. A., Omar, A. F., El-Sebaity, D. M., & Abdelazeem, K. (2021). Pix2pix conditional generative adversarial networks for scheimpflug camera color-coded corneal tomography image generation. *Translational Vision Science and Technology, 10*(7), 21–21.

7. Sandfort, V., Yan, K., Pickhardt, P. J., & Summers, R. M. (2019). Data augmentation using generative adversarial networks (CycleGAN) to improve generalizability in CT segmentation tasks. *Scientific reports, 9*(1), 16884.
8. Kaneko, T., & Kameoka, H. (2018, September). Cyclegan-vc: Non-parallel voice conversion using cycle-consistent adversarial networks. In *2018 26th European signal processing conference (EUSIPCO)* (pp. 2100–2104). IEEE.
9. Torbunov, D., et al. (2023). Uvcgan: Unet vision transformer cycle-consistent gan for unpaired image-to-image translation. In *Proceedings of the IEEE/CVF winter conference on applications of computer vision* (pp. 702–712).
10. Liu, S. (2022, December). Study for identity losses in image-to-image domain translation with cycle-consistent generative adversarial network. *Journal of Physics: Conference Series, 2400*(1), 012030.
11. Hooftman, D., Ziabari, S. S. M., & Snijder, J. (2023, November). Exploring CycleGAN for Bias Reduction in Gender Classification: Generative Modelling for Diversifying Data Augmentation. In *Asian Conference on Pattern Recognition* (pp. 26–40). Cham: Springer Nature Switzerland.
12. Yadav, T., & Sachdeo, R. (2023). Development of Optimal Hyperparameter Tuning-Cycle GAN for Photo-realistic Face Age Progression Model. *International Journal on Artificial Intelligence Tools, 32*(07), 2350068.
13. Liu, B., & Chen, J. (2021). A super resolution algorithm based on attention mechanism and srgan network. *IEEE Access, 9*, 139138–139145.
14. Abbas, R., & Gu, N. (2023). Improving deep learning-based image super-resolution with residual learning and perceptual loss using SRGAN model. *Soft Computing, 27*(21), 16041–16057.
15. Pathak, H. N., Li, X., Minaee, S., & Cowan, B. (2018, December). Efficient super resolution for large-scale images using attentional GAN. In *2018 IEEE International Conference on Big Data (Big Data)* (pp. 1777–1786). IEEE.
16. Lin, J., Xia, Y., Qin, T., Chen, Z., & Liu, T. Y. (2018). Conditional image-to-image translation. In *Proceedings of the IEEE conference on computer vision and pattern recognition* (pp. 5524–5532).
17. Zhang, L., Ji, Y., Lin, X., & Liu, C. (2017, November). Style transfer for anime sketches with enhanced residual u-net and auxiliary classifier gan. In *2017 4th IAPR Asian conference on pattern recognition (ACPR)* (pp. 506–511). IEEE.
18. Azadi, S., Fisher, M., Kim, V. G., Wang, Z., Shechtman, E., & Darrell, T. (2018). Multi-content gan for few-shot font style transfer. In *Proceedings of the IEEE conference on computer vision and pattern recognition* (pp. 7564–7573).
19. Al-Mekhlafi, H., & Liu, S. (2024, September). Artistic style transfer based on attention with knowledge distillation. In *Computer graphics forum* (Vol. 43, No. 6, p. e15127).
20. Wang, H., Wu, P., Rosa, K. D., Wang, C., & Shrivastava, A. (2024). Multimodality-guided Image Style Transfer using Cross-modal GAN Inversion. In *Proceedings of the IEEE/CVF winter conference on applications of computer vision* (pp. 4976–4985).
21. Li, X., et al. (2021). Image-to-image translation via hierarchical style disentanglement. In *Proceedings of the IEEE/CVF conference on computer vision and pattern recognition* (pp. 8639–8648).
22. Shocher, A., Gandelsman, Y., Mosseri, I., Yarom, M., Irani, M., Freeman, W. T., & Dekel, T. (2020). Semantic pyramid for image generation. In *Proceedings of the IEEE/CVF conference on computer vision and pattern recognition* (pp. 7457–7466).
23. Cai, X., et al. (2024). Perceptual loss guided Generative adversarial network for saliency detection. *Information Sciences, 654*, 119625.
24. Satchidanandam, A., Al Ansari, R. M. S., Sreenivasulu, A. L., Rao, V. S., Godla, S. R., & Kaur, C. (2023). Enhancing style transfer with GANs: Perceptual loss and semantic segmentation. *International Journal of Advanced Computer Science and Applications, 14*(11).
25. Laxmi, V., Jagruthi, H., Nallamalli, S. C., & Tiwari, A. (2024). Intelligent art transformation using conditional GANs for image classification, sketch-to-image synthesis and object detection. In *Computer Science Engineering* (pp. 725–734). CRC Press.

26. Shen, Y., Liang, J., & Lin, M. C. (2020). Gan-based garment generation using sewing pattern images. In *Computer Vision–ECCV 2020: 16th European Conference, Glasgow, UK, August 23–28, 2020, Proceedings, Part XVIII 16* (pp. 225–247). Springer International Publishing.

27. Cui, M., Kim, M., Choi, S., & Lee, S. (2022). The usage and impact of GAN in graphic design. *Archives of Design Research, 35*(4), 285–307.

28. Estrada, D. C., Dalgleish, F. R., Den Ouden, C. J., Ramos, B., Li, Y., & Ouyang, B. (2022). Underwater LiDAR image enhancement using a GAN based machine learning technique. *IEEE Sensors Journal, 22*(5), 4438–4451.

29. Moon, C., Uh, Y., & Byun, H. (2018). Image Restoration using GAN. *Journal of Broadcast Engineering, 23*(4), 503–510.

30. Nazeri, K., Ng, E., & Ebrahimi, M. (2018). Image colorization using generative adversarial networks. In *Articulated motion and deformable objects: 10th international conference, AMDO 2018, Palma de Mallorca, Spain, July 12–13, 2018, Proceedings 10* (pp. 85–94). Springer International Publishing.

31. Sushmitha, P., Hanchinamani, G., Madanbhavi, L., & Baligar, V. P. (2023, September). GAN: A Novel Approach for Cartoonizing Real Images. In *2023 Third international conference on ubiquitous computing and intelligent information systems (ICUIS)* (pp. 212–218). IEEE.

32. Zhao, Y., Ren, D., Chen, Y., Jia, W., Wang, R., & Liu, X. (2022). Cartoon image processing: A survey. *International Journal of Computer Vision, 130*(11), 2733–2769.

33. Wang, Y., Gonzalez-Garcia, A., Berga, D., Herranz, L., Khan, F. S., & Weijer, J. V. D. (2020). Minegan: effective knowledge transfer from gans to target domains with few images. In *Proceedings of the IEEE/CVF conference on computer vision and pattern recognition* (pp. 9332–9341).

34. Alotaibi, A. (2020). Deep generative adversarial networks for image-to-image translation: A review. *Symmetry, 12*(10), 1705.

35. Pang, Y., Lin, J., Qin, T., & Chen, Z. (2021). Image-to-image translation: Methods and applications. *IEEE Transactions on Multimedia, 24*, 3859–3881.

36. Kang, M., & Park, J. (2020). Contragan: Contrastive learning for conditional image generation. *Advances in Neural Information Processing Systems, 33*, 21357–21369.

37. Wang, X., Sun, L., Chehri, A., & Song, Y. (2023). A review of GAN-based super-resolution reconstruction for optical remote sensing images. *Remote Sensing, 15*(20), 5062.

38. Elharrouss, O., Almaadeed, N., Al-Maadeed, S., & Akbari, Y. (2020). Image inpainting: A review. *Neural Processing Letters, 51*, 2007–2028.

39. Xiang, H., Zou, Q., Nawaz, M. A., Huang, X., Zhang, F., & Yu, H. (2023). Deep learning for image inpainting: A survey. *Pattern Recognition, 134*, 109046.

40. Dash, A., Wang, G., & Han, T. (2024, May). Attentive partial convolution for RGBD image inpainting. In *Companion proceedings of the ACM on web conference 2024* (pp. 1410–1417).

41. Charef, A., & Ouqour, A. (2024). Improving image inpainting through contextual attention in deep learning. *Engineering, Technology & Applied Science Research, 14*(4), 14904–14909.

42. Dash, A., Gu, J., & Wang, G. (2024). HI-GAN: Hierarchical inpainting GAN with auxiliary inputs for combined rgb and depth inpainting. ArXiv preprint arXiv:2402.10334.

43. Liu, W., Wang, C., & Zhang, Y. (2024). Industrial surface defect detection by multi-scale Inpainting-GAN. *The Visual Computer*, 1–18.

44. Yu, T., et al. (2020, April). Region normalization for image inpainting. In *Proceedings of the AAAI conference on artificial intelligence* (Vol. 34, No. 07, pp. 12733–12740).

45. Kumar, T., Brennan, R., Mileo, A., & Bendechache, M. (2024). Image data augmentation approaches: A comprehensive survey and future directions. *IEEE Access.*

Chapter 3
Autoencoders for Image Augmentation

Autoencoders (AEs) are a class of artificial neural networks designed to learn efficient representations of input data. They consist of an encoder that maps input data to a lower-dimensional latent space, and a decoder that reconstructs the input data from this latent representation. In the context of image augmentation, autoencoders can be used to generate new, yet similar, images [1]. By training an autoencoder on a dataset of images, the network learns to capture the underlying structure and variations within the data. Once trained, the encoder can be used to generate new latent codes, which can then be fed into the decoder to produce augmented images.

3.1 Convolutional AEs

Convolutional AEs (CAEs) excel at image augmentation due to their convolutional layers. These layers effectively capture spatial information in images. CAEs can learn to reconstruct the spatial arrangement of pixels, ensuring that generated augmentations maintain the overall structure of the original image. Convolutional operations are well-suited for learning and reproducing textural patterns, leading to more realistic augmentations [2, 3]. CAEs can capture local variations in brightness, color, and contrast, allowing them to generate diverse augmentations with realistic image details. The key components of autoencoders are encoder and decoder which are discussed in the following subsections.

© The Author(s), under exclusive license to Springer Nature Singapore Pte Ltd. 2025   59
J. Chaki, *The Art of Deep Learning Image Augmentation: The Seeds of Success*,
SpringerBriefs in Computational Intelligence,
https://doi.org/10.1007/978-981-96-5081-1_3

3.1.1 Encoder

The encoder component of an autoencoder is responsible for compressing the input data into a lower-dimensional latent space representation. This compressed representation captures the essential features of the input data while discarding redundant information. A typical encoder for image augmentation comprises a series of convolutional and pooling layers. The initial convolutional layers extract low-level features like textures and edges, while subsequent layers capture higher-level abstractions. Convolutional layers employ filters to convolve over the input image, producing feature maps that highlight specific patterns. The use of the ReLU activation function introduces non-linearity, enabling the network to learn complex representations. Pooling layers, such as max pooling or average pooling, downsample the feature maps, dropping dimensionality while preserving essential information. This hierarchical structure allows the encoder to progressively compress the input image into a lower-dimensional latent space, capturing the most salient features for subsequent reconstruction or manipulation. The number of layers, filter sizes, and pooling configurations can vary depending on the specific application and desired level of compression. The following is an example of simple encoder architecture for a small-sized image:

- Convolutional layer with 32 filters, kernel size 3×3, stride 1, padding 1.
- ReLU activation.
- Max pooling with kernel size 2×2, stride 2.
- Convolutional layer with 64 filters, kernel size 3×3, stride 1, padding 1.
- ReLU activation.
- Max pooling with kernel size 2×2, stride 2
- A fully connected layer to produce the latent code.

3.1.2 Decoder

The decoder in an autoencoder is responsible for reconstructing the original input data from the compressed latent space representation generated by the encoder. It essentially reverses the encoding process. Typically, the decoder mirrors the encoder's architecture but in reverse order. It consists of a series of upsampling and convolutional layers to gradually increase the spatial dimensions of the feature maps. Here upsampling layers increase the spatial dimensions of the feature maps. Techniques like transposed convolution or bilinear interpolation can be used. The convolutional layers refine the upsampled features to produce the final output image. Non-linear activation functions (e.g., ReLU) are applied to introduce non-linearity. The decoder's ability to accurately reconstruct the image is essential for training the autoencoder and for generating new images based on the learned latent space representation. The general idea is to apply a series of linear transformations and non-linear activations to the latent code to reconstruct the image which can be mathematically represented by using the Eq. (3.1).

$$\hat{x} = g(W \times h + b), \tag{3.1}$$

where $\hat{x}$ is the reconstructed image, h is the latent code, W is the weight matrix of the first convolutional layer in the decoder, b is the bias term, g represents the series of convolutional, upsampling, and activation layers in the decoder.

For a 32×32 pixel image, a decoder might consist of:

- Fully connected layer to produce a $4 \times 4 \times 512$ feature map.
- Upsampling layer to produce an $8 \times 8 \times 256$ feature map.
- Convolutional layer with 128 filters, kernel size 3×3, stride 1, padding 1.
- ReLU activation.
- Upsampling layer to produce a $16 \times 16 \times 128$ feature map.
- Convolutional layer with 64 filters, kernel size 3×3, stride 1, padding 1.
- ReLU activation.
- Upsampling layer to produce a $32 \times 32 \times 64$ feature map.
- Convolutional layer with 3 filters (for RGB), kernel size 3×3, stride 1, padding 1.
- Sigmoid activation (or other output activation) to produce the final image.

This is a simplified example, and more complex architectures can be used depending on the desired image size, resolution, and complexity.

3.1.3 Loss Functions in Autoencoders for Image Augmentation

The reconstruction loss is the cornerstone of training an autoencoder. It quantifies the difference between the original input image and the reconstructed output generated by the decoder. This loss function guides the model to minimize the discrepancy between these two images, thereby improving its ability to capture and represent the essential features of the input data [4]. Common choices include:

Mean-Squared Error (MSE): This is the most widely used loss function for image reconstruction [5]. It calculates the average squared difference between the pixel values of the original and reconstructed images as shown in Eq. (3.2).

$$\text{MSE} = \left(\frac{1}{N}\right) \times \sum (x_i - y_i)^2, \tag{3.2}$$

where: N is the total number of pixels, x_i is the pixel value of the original image at position i, y_i is the pixel value of the reconstructed image at position i.

Binary Cross-Entropy (BCE): Used for binary image data (e.g., black and white images) [6], BCE measures the dissimilarity between the two probability distributions as represented in the Eq. (3.3).

$$BCE = -(y * \log(p) + (1 - y) * \log(1 - p)), \tag{3.3}$$

where y is the true pixel value (0 or 1), p is the predicted probability of the pixel being 1.

This equation computes the cross-entropy between the true binary label and the predicted probability for each pixel. The loss is minimized during training to encourage the model to produce accurate pixel predictions.

Categorical Cross-Entropy: Categorical cross-entropy loss is often used in image reconstruction tasks when the pixel values are represented as probabilities or one-hot encoded vectors [7]. This loss function measures the difference between the predicted probability distribution and the true distribution (ground truth) as represented in Eq. (3.4).

$$L = -\sum \left[y_i \times \log(p_i) \right], \tag{3.4}$$

where L is the categorical cross-entropy loss, y_i is the true probability (ground truth) of the i-th class, p_i is the predicted probability of the i-th class, and the summation is over all classes.

For image reconstruction, each pixel can be treated as a multi-class classification problem where the number of classes is equal to the number of possible pixel values (e.g., 256 for 8-bit grayscale images). The ground truth is the pixel value itself, epitomized as a one-hot encoded vector. The model predicts a probability distribution over these classes for each pixel. The categorical cross-entropy loss measures the difference between the predicted and ground truth distributions for all pixels in the image.

3.1.4 Training Process

Autoencoder training involves an iterative process of optimizing the encoder and decoder components [8]. Initially, a dataset of images is collected and preprocessed, followed by random initialization of model parameters. Subsequently, an input image is fed through the encoder to obtain a latent code representation, which is then decoded to reconstruct the original image. The reconstruction error, often calculated using mean squared error or cross-entropy loss, is backpropagated to update the model parameters. Additional loss terms like style loss or perceptual loss can be incorporated for enhanced image quality. This process is repeated for multiple epochs or until convergence, resulting in a trained autoencoder capable of generating similar images to the input data. Using appropriate batch sizes can improve training efficiency and stability. The learning rate determines the step size during optimization. It should be carefully tuned to avoid overshooting or slow convergence. Techniques like dropout, weight decay, or batch normalization can aid in avoiding overfitting and enhancing generalization. We should monitor the validation loss to prevent overfitting and stop training when performance starts to degrade. By iteratively optimizing the autoencoder's parameters, the model learns to effectively capture the underlying structure of the input data and generate high-quality image augmentations.

3.1.5 Improving Reconstruction Quality in CAEs

Several methods can be employed to improve the reconstruction quality of autoencoders. Few of them are mentioned here.

3.1.5.1 Deeper Architectures

Deeper autoencoder architectures have shown significant improvements in image augmentation capabilities [9]. By increasing the number of layers in both the encoder and decoder, these models can capture more complex and intricate features within the image data. Deeper networks can learn more abstract and high-level features, leading to better image reconstruction and generation. Deeper architectures have the potential to generate a wider range of augmented images, including more complex transformations. Deeper models can better handle variations in image style, lighting conditions, and other factors, resulting in more realistic augmentations. Despite the advantages of deeper architecture, a few challenges are there when the architecture becomes deep. Deeper networks are more inclined to overfitting, requiring regularization methods like weight decay or dropout. Training deeper autoencoders can be computationally expensive due to the increased number of parameters. Incorporating residual connections can help alleviate the vanishing gradient problem and improve training stability. Dense connections between layers can enhance information flow and feature reuse. Breaking down the image into multiple scales can capture both fine-grained and coarse-grained features.

3.1.5.2 Skip Connections in Autoencoders

Skip connections are a critical architectural component in autoencoders, particularly for image reconstruction tasks [10]. They establish direct connections between corresponding layers in the encoder and decoder, preserving detailed information that might otherwise be lost during the compression and reconstruction process. Skip connections directly connect the output of one layer to the input of another layer, bypassing intermediate layers. The output of the skip connection is typically concatenated or summed with the output of the main path. While the exact implementation can vary, the core idea of skip connections involves adding the output of an encoder layer to the corresponding decoder layer. Let x be the input to an encoder layer, y be the output of that layer, and z be the output of the corresponding decoder layer. The skip connection can be expressed by using Eq. (3.5).

$$\text{output} = z + x \qquad (3.5)$$

This combined output is then passed to the next layer in the decoder.

There are some benefits of including skip connections in autoencoders for image augmentation. By directly connecting the encoder and decoder layers, skip connections help preserve fine-grained details such as edges, textures, and local patterns. Skip connections facilitate the flow of gradients during backpropagation, preventing the vanishing gradient problem and enabling the training of deeper networks. The combination of low-level details from skip connections and high-level features learned by the encoder and decoder leads to more accurate and visually pleasing reconstructions. U-Net architecture extensively utilizes skip connections to combine feature maps from the encoder with corresponding layers in the decoder. DenseNet can also be adapted for autoencoders by incorporating skip connections between all layers.

3.1.5.3   Residual Connections in Autoencoders

Residual connections, inspired by ResNet architectures, have meaningfully enhanced the performance of autoencoders, especially in image reconstruction tasks [11]. They allow the network to learn residual mappings, making it easier to optimize for identity functions. Residual connections add the input of a layer to the output of that layer. The network learns to optimize a residual function rather than the entire mapping. Let x be the input to a layer, $F(x)$ be the output of the layer without the residual connection, and y be the output with the residual connection. Then:

$$y = F(x) + x \qquad (3.6)$$

The output of the layer is the sum of the original input and the output of the function F.

There are some benefits of including residual connections in autoencoders for image augmentation. Residual connections help to alleviate the vanishing gradient problem, allowing for the training of deeper networks. By adding the original input to the output, residual connections help to preserve fine-grained details in the reconstructed image. Residual connections can speed up the training process by making the optimization landscape smoother. By incorporating residual connections into the encoder and decoder of an autoencoder, it is possible to achieve better reconstruction quality and train deeper models.

3.1.5.4 Perceptual Loss in Autoencoder

Perceptual loss is a critical component in enhancing the perceptual quality of reconstructed images in autoencoders [12]. Unlike pixel-wise loss functions (like MSE), perceptual loss focuses on higher-level image features. A pre-trained image classification network (e.g., VGG) is used to extract feature maps from both the original and reconstructed images. The corresponding feature maps from both images are compared using a loss function like L1 or L2 distance. The perceptual loss is computed as the sum of feature-wise losses across multiple layers of the pre-trained network. Mathematically this can be represented by using Eq. (3.7).

$$L_{\text{perceptual}} = \Sigma_l ||\phi_l(x) - \phi_l(y)||_2^2 \tag{3.7}$$

where $L_{\text{perceptual}}$ is the perceptual loss, $\phi_l(x)$ and $\phi_l(y)$ are the feature maps of the original and reconstructed images at layer l of the pre-trained network, $||.||_2$ is the L2 norm.

There are some benefits of including perceptual loss in autoencoders for image augmentation. By focusing on high-level features, perceptual loss helps in generating images that are perceptually similar to the original. Perceptual loss can help to reduce blurriness in reconstructed images. It can help to preserve fine details and textures in the generated images.

3.1.5.5 Style Loss in Autoencoder

Style loss is a technique used to encourage the generated image to match the style of the original image [13]. It is often used in conjunction with perceptual loss to capture both the content and style of the input image. A pre-trained convolutional neural network (CNN), such as VGG, is used to extract feature maps from both the original and reconstructed images. Gram matrices are computed for corresponding feature maps of the original and reconstructed images. The Gram matrix captures the style information by measuring the correlation between different feature channels. The style loss is computed as the difference between the Gram matrices of the original and reconstructed images. Mathematically this can be represented by using Eq. (3.8).

$$L_{\text{style}} = \Sigma_l ||G_l^x - G_l^y||_F^2, \tag{3.8}$$

where L_{style} is the style loss, G_l^x and G_l^y are the Gram matrices of the original and reconstructed images at layer l, $||.||_F$ is the Frobenius norm.

There are some benefits of including style loss in autoencoders for image augmentation. Style loss helps to maintain the overall style and appearance of the original image, such as color palette, texture, and brushstrokes. By combining style loss with perceptual loss, it is possible to generate augmented images that not only preserve the content but also maintain the aesthetic qualities of the original image. By incorporating style loss into the autoencoder training process, it is possible to generate augmented images that are not only similar in content but also visually appealing.

3.1.5.6 Dropout in Autoencoders for Image Augmentation

Dropout is a regularization technique that helps prevent overfitting in neural networks, including autoencoders [14]. It involves randomly setting a fraction of input units to zero at each training update. During training, each neuron has a probability p of being included in the network. The remaining neurons are temporarily dropped out. This process is equivalent to training a large number of neural networks with different architectures. By applying dropout to the hidden layers of the encoder, we encourage the network to learn more robust features that are not overly reliant on specific neurons. Dropout can also be applied to the decoder layers to prevent overfitting and improve generalization. Let x be the input to a neuron, and p be the dropout probability. The output y with dropout applied is represented by using Eq. (3.9).

$$y = x \times (1 - p), \tag{3.9}$$

where $1 - p$ is the probability of a neuron being kept.

Dropout reduces the network's reliance on specific neurons, making it more robust to noise and variations in the input data. By forcing the network to learn multiple representations, dropout enhances its ability to generalize to unseen data. Dropout can introduce randomness in the reconstruction process, leading to more diverse and realistic augmented images. The choice of dropout rate is crucial. A higher dropout rate can lead to underfitting, while a lower rate might not be effective in preventing overfitting. To simplify testing, it's common to scale the activations by $1/(1 - p)$ during training. This allows for using the same weights during testing without the need for dropout. By incorporating dropout into the autoencoder architecture, it is possible to improve the model's generalization ability and produce higher-quality image augmentations.

3.1.5.7 Weight Decay for Autoencoder Image Augmentation

Weight decay, also known as L2 regularization, is a technique used to prevent overfitting in neural networks, including autoencoders [15]. It penalizes large weights in the network, encouraging the model to learn more generalized representations. The weight decay term is added to the loss function as represented in Eq. (3.10).

$$L_{\text{total}} = L_{\text{reconstruction}} + \lambda \times L_{\text{decay}}, \tag{3.10}$$

where L_{total} is the total loss, $L_{reconstruction}$ is the reconstruction loss (e.g., MSE, cross-entropy), L_{decay} is the weight decay term, λ is the regularization parameter, controlling the strength of the penalty.

The weight decay term is typically calculated as the sum of the squared weights in the network as represented in the Eq. (3.11).

$$L_{decay} = \Sigma\left(w_i^2\right), \tag{3.11}$$

where w_i is the weight of the i-th neuron.

By penalizing large weights, weight decay discourages the model from memorizing the training data. A model with smaller weights is more likely to generalize well to unseen data. By preventing overfitting, weight decay can help the autoencoder generate more diverse and realistic augmented images. Weight decay can be easily implemented in most deep-learning frameworks by adding a regularization term to the loss function. The regularization parameter λ needs to be tuned appropriately to balance the trade-off between model complexity and generalization performance. By incorporating weight decay, autoencoders can produce more robust and generalizable image augmentations.

3.1.5.8 Batch Normalization for Autoencoder Image Augmentation

Batch normalization is a technique that normalizes the activations of a layer across a mini-batch of training examples [16]. This helps stabilize the training procedure and enhance the model's performance. For a given layer with activations x, the batch normalization process involves the following steps.

Calculate the mean and variance of the activations as represented in Eq. (3.12).

$$\begin{aligned} \mu_B &= E[x] \\ \sigma_B^2 &= \mathrm{Var}[x], \end{aligned} \tag{3.12}$$

where μ_B is the mean of the batch, and σ_B^2 is the variance of the batch.
Normalize the activations as shown in the Eq. (3.13).

$$\hat{x} = \frac{(x - \mu_B)}{\sqrt{\left(\sigma_B^2 + \varepsilon\right)}}, \tag{3.13}$$

where ε is a small constant to prevent division by zero.
Scale and shift as shown in the Eq. (3.14).

$$y = \gamma \times \hat{x} + \beta, \tag{3.14}$$

where γ and β are learnable parameters that allow the network to rescale and shift the normalized activations.

By normalizing the activations, batch normalization helps to stabilize the training process and improve convergence. Batch normalization can aid avoid overfitting by reducing the sensitivity of the network to the specific distribution of the training data. Batch normalization can accelerate training by allowing the use of higher learning rates. By incorporating batch normalization into the autoencoder architecture, it is possible to achieve faster convergence, better generalization, and improved image augmentation quality.

3.1.5.9 Latent Space Interpolation for Autoencoder Image Augmentation

Latent space interpolation is a powerful technique for generating new, intermediate images using autoencoders [17]. By manipulating the latent code representations, it's possible to create smooth transitions between different images. Two input images are encoded into their respective latent space representations, z_1 and z_2. A linear interpolation is performed between z_1 and z_2 to generate intermediate latent codes as represented in the Eq. (3.15).

$$z_{\text{interp}} = \alpha \times z_1 + (1 - \alpha) \times z_2, \tag{3.15}$$

where α is a scalar value between 0 and 1, controlling the interpolation weight.

Each interpolated latent code Z_{interp} is decoded into a corresponding image using the decoder.

This technique generates intermediate images with gradual changes in appearance. It creates morphing effects between different images. It generates new training samples by interpolating between existing data points.

3.1.5.10 Latent Space Arithmetic for Autoencoder Image Augmentation

Latent space arithmetic involves performing mathematical operations directly on the latent code representations of images [18]. This technique can lead to interesting and creative image manipulations.

Let z_1 and z_2 be the latent code representations of two images. Basic arithmetic operations can be performed on these codes:

Addition: $z_{\text{new}} = z_1 + z_2$.

Subtraction: $z_{\text{new}} = z_1 - z_2$.

Scaling: $z_{\text{new}} = \alpha \times z_1$ where α is a scalar value.

Combination: $z_{\text{new}} = \alpha \times z_1 + (1 - \alpha) \times z_2$ (weighted combination).

The resulting latent code z_{new} can then be decoded to generate a new image.

By combining the latent codes of two images, new images with blended features can be created. This method creates smooth transitions between images by linearly interpolating their latent codes. We should explore regions of the latent space beyond the training data to generate novel images. By manipulating specific dimensions of the latent code, it's possible to transfer styles between images. By carefully exploring the latent space and combining latent space arithmetic with other image manipulation techniques, it's possible to generate diverse and creative image augmentations.

3.2 Denoising Autoencoders

Denoising autoencoders (DAEs) are a variant of autoencoders specifically trained to remove noise from corrupted input images [19]. By learning to reconstruct clean images from noisy versions, DAEs implicitly learn robust feature representations. DAEs excel at learning robust features by being trained on noisy inputs. This makes them well-suited for generating diverse and realistic augmentations, as they can handle various image corruptions. By forcing the model to reconstruct clean images from noisy inputs, DAEs develop a strong ability to ignore noise and focus on essential image information.

This autoencoder is trained on noisy versions of clean images. Noise (e.g., Gaussian noise) is added to the original image to create a noisy input for the autoencoder. The noisy input image is fed into the encoder, which maps it to a lower-dimensional latent space representation. The latent code is passed through the decoder to reconstruct the original clean image. A loss function, typically mean squared error (MSE), is used to compare the reconstructed image with the original, clean image. The goal is to minimize this loss. Through this process, the autoencoder learns to extract underlying features that are resistant to noise, enabling it to generate diverse and clean augmentations. Let: x be the original clean image, x' be the noisy input image ($x +$ noise), h be the latent code representation and $\hat{x}$ be the reconstructed image.

The encoder can be represented as shown in the Eq. (3.16).

$$h = f(x'), \tag{3.16}$$

where f is the encoder function.

The decoder can be represented as shown in the Eq. (3.17).

$$\hat{x} = g(h), \tag{3.17}$$

where g is the decoder function.

The loss function (MSE) is shown in the Eq. (3.18).

$$L = ||x - \hat{x}||^2 \tag{3.18}$$

The goal is to minimize L through backpropagation.

By adding different types of noise (Gaussian, salt-and-pepper, etc.) to the original image, the autoencoder learns to handle various image corruptions, leading to robust augmentations. Randomly erasing image patches can be seen as a form of noise

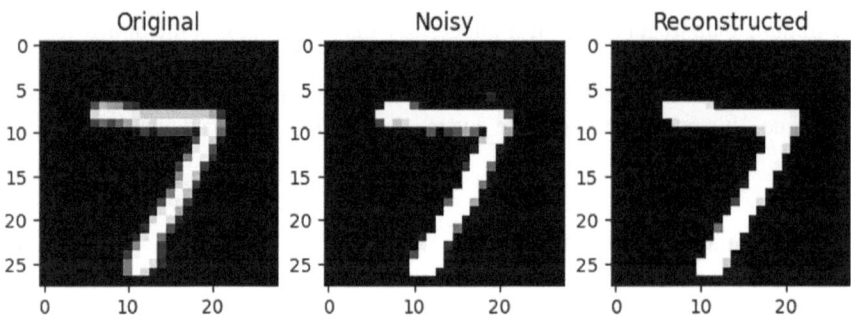

Fig. 3.1 The output by using DAE on the MNIST dataset

and can be handled by DAEs. Introducing JPEG compression artifacts can create challenging scenarios for the autoencoder, leading to more robust augmentations. Denoising autoencoders can be utilized for several image-processing tasks beyond denoising, like image inpainting and super-resolution. By training a DAE on a dataset of noisy images, the model learns to effectively remove noise and generate clean, high-quality outputs. This makes DAEs a valuable tool for image augmentation and other image-processing applications. Figure 3.1 shows the output by using DAE on the MNIST dataset.

3.3 Variational Autoencoders

Variational autoencoders (VAEs) are generative models that learn a probabilistic representation of the input data [20]. Unlike traditional autoencoders, VAEs introduce a latent space that encodes the underlying representation of the image. By sampling from this space, new image variations can be generated, leading to diverse augmentations. VAEs learn a probability distribution in the latent space, allowing for more controlled exploration and generation of new images. A VAE consists of two main components: an encoder and a decoder. The encoder maps input data x to the parameters of a latent space distribution, typically a Gaussian. It outputs the mean μ and variance σ of the latent space distribution. The encoder learns to map input data points to regions of high probability density in the latent space. The decoder takes a sample z from the latent space distribution and reconstructs the input data x. The decoder learns to map points in the latent space back to the original data space.

The VAE loss function consists of two terms:

Reconstruction Loss: Measures the difference between the original input x and the reconstructed output x̂. Typically, mean squared error (MSE) is used.

$$L_{rec} = ||x - \hat{x}||^2 \tag{3.19}$$

KL Divergence Loss: KL Divergence (KLD) is a measure of how one probability distribution diverges from a second, reference probability distribution. In the context of variational autoencoders (VAEs), it's used to ensure the latent space distribution approximates a prior distribution, typically a standard normal distribution as represented in the Eq. (3.20).

$$L_{KL} = -0.5 \times \sum \left(1 + \log\left(\sigma^2\right) - \mu^2 - \sigma^2\right), \tag{3.20}$$

where the sum is over all dimensions of the latent space.

The total loss is the sum of the reconstruction loss and the KL divergence loss as shown in the Eq. (3.21).

$$L_{\text{total}} = L_{\text{rec}} + \beta \times L_{KL} \tag{3.21}$$

where β is a hyperparameter balancing the two terms.

While training, the reparameterization trick technique stabilizes training by sampling from a standard normal distribution and scaling the sample by the standard deviation of the latent distribution. Using specific initialization schemes, e.g., Kingma and Welling Initialization can improve training convergence. Also, gradually decreasing the learning rate during training can help to refine the model. Carefully tuning hyperparameters such as the learning rate, batch size, and latent space dimensionality is crucial for optimal performance. Experimenting with different VAE architectures and loss functions can aid find the finest model for a specific task. Figure 3.2 shows some handwritten image samples generated using VAE (the model was trained using the MNIST dataset).

To generate new images, a random sample is drawn from the prior distribution (usually a standard normal distribution) in the latent space. This sample is then fed into the decoder to produce a new image. The latent space is continuous, allowing for smooth interpolation and manipulation of images. The KL divergence term acts as a regularizer, preventing the model from collapsing to deterministic solutions. By incorporating probabilistic reasoning, VAEs provide a more flexible and powerful approach to image augmentation compared to traditional autoencoders. Combining VAEs with GANs can enhance the quality of generated samples.

3.4 Adversarial Autoencoders

Adversarial Autoencoders (AAEs) combine the principles of autoencoders and generative adversarial networks (GANs) to improve image generation and augmentation [21]. This approach aims to learn a better latent space representation by introducing adversarial training. An AAE consists of three components. The encoder maps input images to a latent space distribution. The decoder reconstructs images from the latent space and the discriminator distinguishes between real and reconstructed images. The

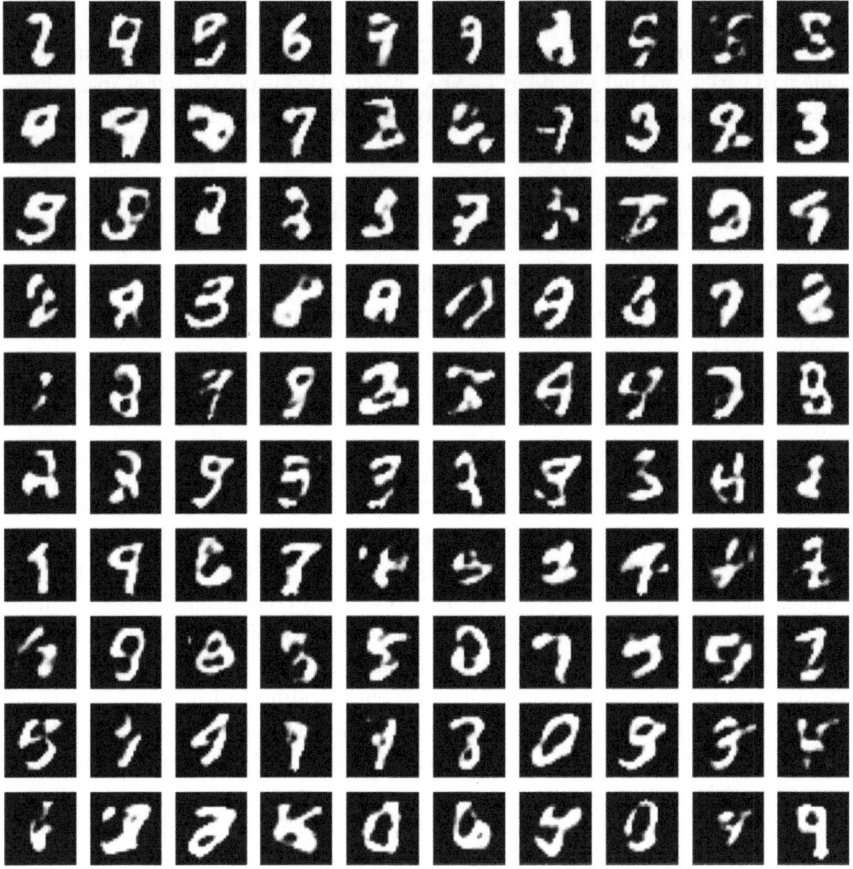

Fig. 3.2 Handwritten image samples generated using VAE

AAE's loss function combines the reconstruction loss (discussed in previous sections) of a standard autoencoder with an adversarial loss (discussed in previous chapters). The AAE is trained in an adversarial manner. The generator aims to produce images that can fool the discriminator, while the discriminator tries to distinguish between real and fake images. This competitive process leads to the generation of more realistic and diverse images. By introducing adversarial training, AAEs can generate higher-quality images compared to standard autoencoders. The adversarial component encourages the model to explore different regions of the latent space, leading to more diverse image augmentations. The adversarial loss helps to learn a more meaningful latent space representation. By combining the reconstruction capabilities of autoencoders with the discriminative power of GANs, AAEs offer a promising approach to image augmentation. Figure 3.3 shows some handwritten image samples generated using AAE (the model was trained using the MNIST dataset).

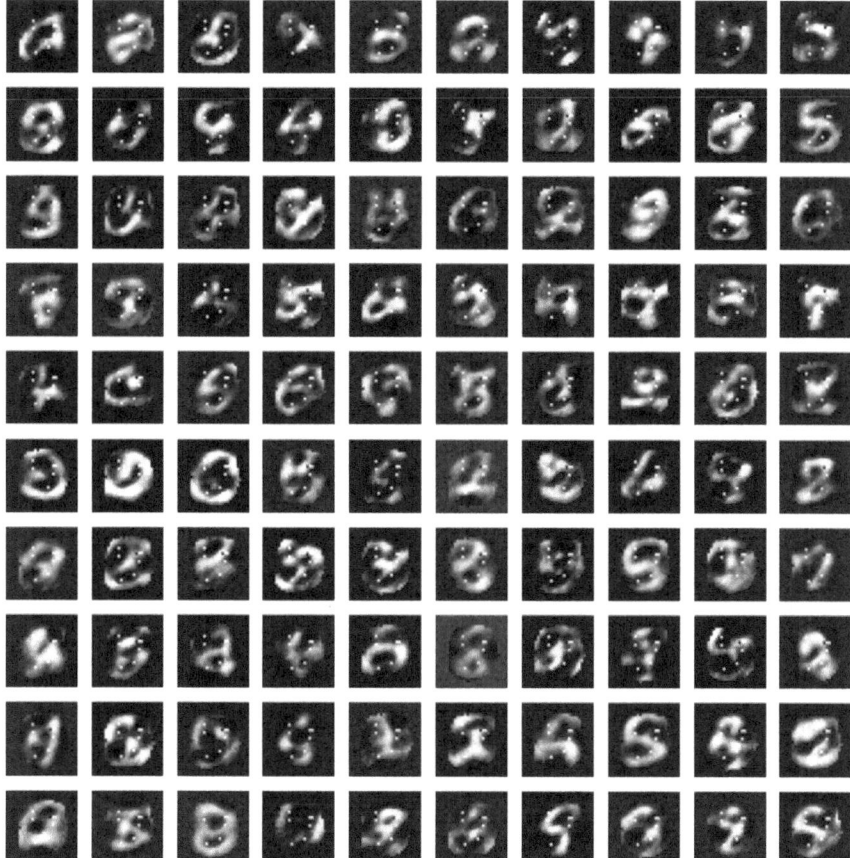

Fig. 3.3 Handwritten image samples generated using AAE

3.5 Applications of Different Autoencoders in Image Augmentation

Autoencoders can significantly augment training datasets by generating synthetic samples that resemble the original data distribution. This is particularly beneficial in scenarios with limited data, imbalanced classes, or a need for diverse training examples. An autoencoder is trained on the available dataset to learn a compressed representation of the data. By sampling random points in the latent space and decoding them, new synthetic data points can be created. The generated samples are added to the original training dataset to increase its size and diversity.

In medical imaging, diseases like certain types of cancer often have limited data available for training models [22]. Autoencoders can generate synthetic images of these rare diseases, improving model performance and aiding in early detection. Autoencoders can be instrumental in creating synthetic medical images that reserve the statistical properties of original data while safeguarding patient privacy. This approach offers a promising solution to the challenges posed by sharing sensitive medical information. By augmenting the dataset with autoencoder-generated images, object detection models can become more robust to variations in object appearance, lighting conditions, and occlusions. Autoencoder-generated images can be used as additional training data for GANs, enhancing the quality and diversity of generated images. Autoencoders can extract style information from images, which can be transferred to other images using techniques like style transfer. Autoencoders can be used to reconstruct normal data points, and deviations from the reconstruction error can indicate anomalies. Autoencoders can be applied to image compression by learning efficient representations of image data. While not as specialized as super-resolution models, autoencoders can be trained to enhance low-resolution images, providing additional training data.

By expanding the training dataset with synthetic images generated by autoencoders, machine-learning models can achieve higher accuracy, better generalization, and improved robustness.

3.6 Challenges and Future Scopes of Autoencoders in Image Augmentation

Autoencoder performance hinges on careful architecture design and hyperparameter selection. Overfitting happens when an autoencoder learns the training data too well, capturing irrelevant details and noise [23]. This causes poor generalization and limited augmentation diversity. Regularization techniques like dropout, weight decay, and early stopping can help prevent overfitting. Underfitting happens when the model is too simple to capture the underlying data distribution. This results in low-quality augmentations and poor reconstruction performance. Increasing model complexity by adding more layers, and neurons, or using more complex activation functions can help address underfitting. A larger latent space can capture more information but increases the risk of overfitting. The depth of the encoder and decoder affects the model's capacity to learn complex patterns. The learning rate controls the step size during optimization. A suitable learning rate is essential for convergence. The number of samples handled in one iteration (batch size) affects the training speed and stability. The choice of optimizer (e.g., Adam, SGD) can impact convergence and performance. The choice of activation functions (e.g., ReLU, LeakyReLU) affects the network's ability to learn non-linear relationships. Balancing the complexity of the encoder and decoder is crucial for effective reconstruction. The latent space dimension should be significantly smaller than the input image size to encourage efficient encoding. Incorporating skip connections can improve information flow and reconstruction quality. By carefully considering these factors and conducting experiments, it's possible to design an autoencoder architecture that effectively generates diverse and realistic image augmentations.

The choice of autoencoder architecture significantly impacts the quality and diversity of generated augmentations [24]. While powerful, CAEs might struggle with abstract or semantic features that are not directly related to the spatial arrangement of pixels. This can lead to limitations in understanding image content. CAEs might not capture high-level semantic information, such as the objects present or the scene depicted in the image. Augmentations might appear visually realistic but lack the semantic context of the original image. While promising, VAEs can suffer from a significant challenge known as mode collapse. During training, the VAE might get stuck in specific regions of the latent space, leading to the generation of very similar images. This limits the diversity of augmentations and hinders the model's capability to explore the full range of image variations. If the latent space dimensionality is too small, it might not be able to capture the full complexity of the data distribution, causing mode collapse. The KL divergence term in the VAE loss function encourages diversity. An inappropriate weight given to this term can result in either mode collapse or neglecting valuable information in the latent space. Providing a larger latent space offers more space for diverse representations, reducing the chance of mode collapse. The variation of VAE, β-VAE, uses a hyperparameter β to control the influence of the KL divergence term, allowing for a better balance between reconstruction accuracy and latent space diversity. Dropout, weight decay, and other regularization approaches can help prevent VAEs from focusing on specific modes. While DAEs are effective at removing noise and preserving low-level image features, they might struggle to capture high-level semantic information. This can limit their ability to generate semantically meaningful augmentations. The performance of DAEs is influenced by the type of noise used during training. If the test data contains different noise patterns, the model's effectiveness might be reduced.

Autoencoders have shown significant potential in image augmentation. However, several avenues for future research and development remain. To further enhance the capabilities of autoencoders for image augmentation, researchers are exploring hybrid architectures that combine the strengths of different models. Integrating attention mechanisms allows the network to focus on specific image regions, improving feature extraction and reconstruction [25]. Additionally, using self-supervised learning techniques can enhance feature representation learning without the need for explicit labels, making the model more data-efficient. These advancements hold promise for creating even more sophisticated and effective image augmentation techniques.

To achieve more precise and creative image augmentations, researchers are focusing on refining latent space manipulation techniques. Disentangling latent factors enables control over specific image attributes, allowing for targeted modifications [26]. This opens doors for semantic editing, where users can directly manipulate image elements within the latent space. Furthermore, using autoencoder-based latent space representations can enhance style transfer capabilities, enabling the creation of images with diverse artistic styles while preserving content integrity.

To expand the practical utility of autoencoder-based image augmentation, efforts are directed toward developing efficient architectures suitable for real-time processing on resource-constrained devices [27]. This involves optimizing model size and computational complexity without compromising augmentation quality. Additionally, enabling users to interactively control the augmentation process through the latent space empowers users to create tailored image variations, expanding the creative possibilities of this technique.

Developing robust evaluation metrics is crucial for assessing the quality and effectiveness of augmented images generated by autoencoders. Traditional image quality metrics like Peak Signal-to-Noise Ratio (PSNR) and Structural Similarity Index (SSIM) might not fully capture the perceptual quality of augmented images. To address this, researchers are exploring more comprehensive evaluation approaches like diversity metrics which are used to measure the diversity of generated images, such as Fréchet Inception Distance (FID) [28] or Kernel Inception Distance (KID) [29]. FID measures the distance between the feature distributions of real and generated images in a high-dimensional feature space. It's calculated by modeling these distributions as multivariate Gaussian distributions and computing the distance between their means and covariance matrices. A lower FID score indicates higher similarity between real and generated image distributions, implying better diversity. KID is a non-parametric alternative to FID, using kernel methods to estimate the distance between the distributions of real and generated images. It often provides more robust results compared to FID, especially for complex image distributions. Both FID and KID are valuable tools for measuring the diversity of generated images and comparing different image generation models. However, they are not perfect and should be used in conjunction with other metrics for a comprehensive evaluation. Downstream task performance can be used to evaluate the impact of augmented data on the performance of downstream tasks (e.g., classification, object detection) to assess augmentation quality indirectly. Sometimes human evaluation also can be important to incorporate human judgment to assess the visual quality and realism of augmented images. By combining multiple evaluation metrics, a more comprehensive assessment of autoencoder-generated augmentations can be achieved. By exploring these directions, autoencoders can become even more powerful tools for image augmentation and contribute to the advancement of various applications.

3.7 Summary

In this chapter different aspects of autoencoder-based image augmentation are discussed. Autoencoders, neural networks designed to reconstruct input data, have emerged as powerful tools for image augmentation. By learning efficient representations of images, autoencoders can generate diverse and realistic augmentations. Key techniques include VAEs for probabilistic generation, DAEs for handling noisy data, and AAEs for enhancing image quality. Architectural considerations such as depth, skip connections, and attention mechanisms play crucial roles in model performance. However, challenges like overfitting, mode collapse, and computational cost require careful attention. Evaluating the effectiveness of augmentation often involves downstream task performance metrics and diversity metrics like FID and KID.

References

1. Shorten, C., & Khoshgoftaar, T. M. (2019). A survey on image data augmentation for deep learning. *Journal of Big Data, 6*(1), 1–48.
2. Nguyen, T. P., Chae, D. S., Choi, S. H., Jeong, K., & Yoon, J. (2023). Enhancement of hip X-ray with convolutional autoencoder for increasing prediction accuracy of bone mineral density. *Bioengineering, 10*(10), 1169.
3. Shakibania, H., Raoufi, S., & Khotanlou, H. (2025). Cdan: Convolutional dense attention-guided network for low-light image enhancement. *Digital Signal Processing, 156*, 104802.
4. Mujeeb, A., Dai, W., Erdt, M., & Sourin, A. (2018, October). Unsupervised surface defect detection using deep autoencoders and data augmentation. In *2018 International conference on cyberworlds (CW)* (pp. 391–398). IEEE.
5. Ohno, H. (2020). Auto-encoder-based generative models for data augmentation on regression problems. *Soft Computing, 24*(11), 7999–8009.
6. Singh, P., Chen, L., Chen, M., Pan, J., Chukkapalli, R., Chaudhari, S., & Cirrone, J. (2023). Enhancing medical image segmentation: Optimizing cross-entropy weights and post-processing with autoencoders. In *Proceedings of the IEEE/CVF international conference on computer vision* (pp. 2684–2693).
7. Saldanha, J., Chakraborty, S., Patil, S., Kotecha, K., Kumar, S., & Nayyar, A. (2022). Data augmentation using variational autoencoders for improvement of respiratory disease classification. *PLoS ONE, 17*(8), e0266467.
8. Berahmand, K., Daneshfar, F., Salehi, E. S., Li, Y., & Xu, Y. (2024). Autoencoders and their applications in machine learning: A survey. *Artificial Intelligence Review, 57*(2), 28.
9. Yang, Y., Wu, Q. J., & Wang, Y. (2016). Autoencoder with invertible functions for dimension reduction and image reconstruction. *IEEE Transactions on Systems, Man, and Cybernetics: Systems, 48*(7), 1065–1079.
10. Zhao, G., Liu, J., Jiang, J., Guan, H., & Wen, J. R. (2018, August). Skip-connected deep convolutional autoencoder for restoration of document images. In *2018 24th international conference on pattern recognition (ICPR)* (pp. 2935–2940). IEEE.
11. Zini, S., Bianco, S., & Schettini, R. (2020). Deep residual autoencoder for blind universal jpeg restoration. *IEEE Access, 8*, 63283–63294.
12. Pihlgren, G. G., Sandin, F., & Liwicki, M. (2020, July). Improving image autoencoder embeddings with perceptual loss. In *2020 International Joint Conference on Neural Networks (IJCNN)* (pp. 1–7). IEEE.
13. Zhu, Q., Wang, H., & Zhang, R. (2021). Wavelet loss function for auto-encoder. *IEEE Access, 9*, 27101–27108.
14. Srivastava, N., Hinton, G., Krizhevsky, A., Sutskever, I., & Salakhutdinov, R. (2014). Dropout: A simple way to prevent neural networks from overfitting. *The journal of machine learning research, 15*(1), 1929–1958.
15. Moradi, R., Berangi, R., & Minaei, B. (2020). A survey of regularization strategies for deep models. *Artificial Intelligence Review, 53*(6), 3947–3986.
16. Chen, H., Wang, Y. H., & Fan, C. H. (2021). A convolutional autoencoder-based approach with batch normalization for energy disaggregation. *The Journal of Supercomputing, 77*(3), 2961–2978.
17. Liu, X., Zou, Y., Kong, L., Diao, Z., Yan, J., Wang, J., Jia, P., & You, J. (2018, August). Data augmentation via latent space interpolation for image classification. In *2018 24th international conference on pattern recognition (ICPR)* (pp. 728–733). IEEE.
18. Khan, M. F. I., Hossain, Z., Hossen, A., Alam, M. N. U., Masum, A. K. M., & Uddin, M. Z. (2024). High-fidelity reconstruction of 3D temperature fields using attention-augmented CNN autoencoders with optimized Latent Space. *IEEE Access*.
19. Creswell, A., & Bharath, A. A. (2018). Denoising adversarial autoencoders. *IEEE Transactions on Neural Networks and Learning Systems, 30*(4), 968–984.

20. Ivanovic, B., Leung, K., Schmerling, E., & Pavone, M. (2020). Multimodal deep genera-tive models for trajectory prediction: A conditional variational autoencoder approach. *IEEE Robotics and Automation Letters, 6*(2), 295–302.

21. Alqahtani, H., Kavakli-Thorne, M., & Kumar, G. (2021). Applications of generative adversarial networks (gans): An updated review. *Archives of Computational Methods in Engineering, 28*, 525–552.

22. Chlap, P., Min, H., Vandenberg, N., Dowling, J., Holloway, L., & Haworth, A. (2021). A review of medical image data augmentation techniques for deep learning applications. *Journal of Medical Imaging and Radiation Oncology, 65*(5), 545–563.

23. Kumar, T., Brennan, R., Mileo, A., & Bendechache, M. (2024). Image data augmentation approaches: A comprehensive survey and future directions. *IEEE Access*.

24. Mumuni, A., & Mumuni, F. (2022). Data augmentation: A comprehensive survey of modern approaches. *Array, 16*, 100258.

25. Duan, R., Chen, Z., Zhang, H., Wang, X., Meng, W., & Sun, G. (2023). Dual residual denoising autoencoder with channel attention mechanism for modulation of signals. *Sensors, 23*(2), 1023.

26. Jha, A. H., Anand, S., Singh, M., & Veeravasarapu, V. R. (2018). Disentangling factors of variation with cycle-consistent variational auto-encoders. In: *Proceedings of the European conference on computer vision (ECCV)* (pp. 805–820).

27. Kamath, V., & Renuka, A. (2023). Deep learning based object detection for resource constrained devices: Systematic review, future trends and challenges ahead. *Neurocomputing, 531*, 34–60.

28. Buzuti, L. F., & Thomaz, C. E. (2023). Fréchet AutoEncoder distance: A new approach for evaluation of generative adversarial networks. *Computer Vision and Image Understanding, 235*, 103768.

29. Knop, S., Mazur, M., Spurek, P., Tabor, J., & Podolak, I. (2022). Generative models with kernel distance in data space. *Neurocomputing, 487*, 119–129.

Chapter 4
Applications of Deep Learning-Based Image Augmentation

Image augmentation, a foundation in the realm of deep learning, involves artificially expanding a training dataset to enhance model performance and generalization. By creating diverse and realistic variations of existing images, augmentation mitigates the risk of overfitting and enhances the model's ability to handle real-world data complexities [1]. This chapter delves into the applications of deep learning-based image augmentation techniques, exploring their impact on various domains and their potential to revolutionize image-centric tasks. Some of the applications are discussed here.

4.1 Image Detection and Classification

In domains characterized by data scarcity, such as medical imaging or rare object classification, image augmentation emerges as a crucial technique to prevent overfitting and enhance model performance. With insufficient training data, models tend to memorize training examples rather than learning generalizable features. This leads to poor performance on unseen data. Models trained on limited data often exhibit suboptimal performance due to a lack of exposure to diverse image variations. Augmentation artificially increases the size of the training dataset by creating new, synthetic images. By exposing the model to a wider range of image variations, augmentation enhances its ability to generalize to unseen data. Augmentation reduces the risk of overfitting by providing the model with more diverse training examples. By effectively utilizing image augmentation techniques, researchers and practitioners can address the challenges posed by limited datasets and develop more robust and accurate models.

Image augmentation plays a pivotal role in enhancing the robustness of deep learning models to real-world image variations [2–4]. By exposing the model to a diverse range of transformed images, it learns to recognize and classify objects or

© The Author(s), under exclusive license to Springer Nature Singapore Pte Ltd. 2025
J. Chaki, *The Art of Deep Learning Image Augmentation: The Seeds of Success*,
SpringerBriefs in Computational Intelligence,
https://doi.org/10.1007/978-981-96-5081-1_4

patterns irrespective of changes in lighting, scale, orientation, or other image properties. Image augmentation plays a vital role in allowing deep learning models to extract more robust and discriminative features. By exposing the model to a variety of transformed images, it learns to focus on essential image characteristics that are invariant to these transformations. The model learns to extract features that are consistent across different image variations, such as rotations, flips, or scale changes. This leads to a more robust representation of the image content. By encountering diverse image examples, the model is forced to differentiate between subtle variations in image content, leading to the extraction of more discriminative features. By training on rotated images, the model learns to recognize objects regardless of their orientation. Exposure to images at different scales helps the model extract features that are invariant to object size. Adding noise to images forces the model to learn features that are resistant to image degradation. Augmentation helps the model adapt to real-world image conditions, such as changes in lighting, occlusion, or background clutter. Augmenting underrepresented classes can help balance the dataset and improve classification accuracy.

Adversarial attacks pose a significant threat to the reliability of deep learning models, particularly in image classification [5]. These attacks comprise making subtle modifications to input images that can drastically alter model predictions. Image augmentation can play a vital role in improving model robustness against such attacks. By exposing the model to a wide range of augmented images, it learns to extract more robust and discriminative features. This makes it harder for adversaries to manipulate the input image to mislead the model. While image augmentation can improve robustness, combining it with adversarial training can provide even stronger defenses. The level of augmentation should be carefully balanced to avoid overfitting or degrading model performance. The effectiveness of augmentation techniques against evolving adversarial attacks should be continuously monitored and adapted. By incorporating image augmentation into the training process, models can become more resilient to adversarial attacks, ensuring their reliability and security in real-world applications.

Deep learning-based image augmentation has significantly improved the performance of object recognition models. By exposing the model to a variety of transformed images, it learns to detect objects under various real-world conditions. Extracting random crops from images helps the model to focus on different object scales and positions. Scaling images to different sizes exposes the model to objects at various scales. Combining multiple images into a single image can improve the model's ability to detect objects in complex scenes [6]. Randomly occluding parts of objects can make the model more robust to real-world scenarios where objects are partially obscured. Removing random image patches or rectangular regions can simulate occlusions.

Here is a real-time application of image augmentation using deep learning. Breast cancer is a significant global health concern, and early detection is crucial for successful treatment. Image augmentation plays a vital role in refining the accuracy and robustness of breast cancer recognition models [7]. Obtaining a diverse and large dataset of breast cancer images can be challenging due to privacy concerns and the rarity of certain cancer types. Breast images vary significantly in terms of density, size, and orientation, making it difficult for models to generalize. The number of benign and malignant cases is often imbalanced, affecting model performance. Image augmentation techniques like rotation, flipping, scaling, and cropping can simulate different imaging angles and perspectives. Adjustments in brightness, contrast, and color can mimic variations in image acquisition conditions. Introducing Gaussian or salt-and-pepper noise can make the model more robust to image degradation. Simulating real-world tissue deformations can improve model generalization. Using pre-built augmentation libraries (e.g., Albumentations, Imgaug) can streamline the process. By exposing the model to a wider range of image variations, augmentation enhances its ability to detect subtle abnormalities. Augmentation helps prevent over-fitting, especially in data-limited scenarios. Models trained on augmented data are more likely to generalize to unseen images, improving diagnostic accuracy. Synthetic images generated through augmentation can protect patient privacy while maintaining data utility. By effectively applying image augmentation techniques, breast cancer detection models can achieve higher sensitivity, specificity, and accuracy, leading to earlier and more accurate diagnoses.

4.2 Image Segmentation

Image augmentation is a critical component in improving the performance of image segmentation models. By increasing the diversity of training data, augmentation helps models generalize better and achieve higher accuracy.

Preserving pixel-level correspondence between an image and its corresponding mask is crucial for image segmentation tasks [8]. When applying augmentations, it's essential to apply the same transformations to both the image and the mask to maintain this alignment. Augmentations like cropping, scaling, and flipping can help models better detect and segment small objects. For example, consider applying a random rotation to a medical image and its corresponding mask. Both the image and the mask should be rotated by the same angle to maintain the correct spatial relationship between image features and their corresponding labels.

```
import cv2
def rotate_image_and_mask(image, mask, angle):
  # ... (rotation logic)
  rotated_image = cv2.warpAffine(image, M, (width, height))
  rotated_mask = cv2.warpAffine(mask, M, (width, height))
  return rotated_image, rotated_mask
```

By ensuring pixel-level correspondence between the image and mask, the augmented data can be effectively used to train segmentation models without compromising accuracy.

Simulating occlusions during image augmentation is crucial for training robust object detection and segmentation models [9]. By exposing models to images with occluded objects, they become better equipped to handle real-world scenarios where objects are partially or fully obscured. For example, Occluding parts of vehicles, pedestrians, or traffic signs in images can improve the model's ability to detect objects under challenging conditions. Simulating organ occlusions by other organs or tissues can enhance the model's performance in complex medical images. By incorporating occlusion handling techniques into image augmentation pipelines, models can become more resilient to real-world challenges and improve overall performance. While simulating occlusions is valuable for improving model robustness, it also presents several challenges [10]: (1) Preserving Object Integrity: Ensuring that occlusions do not completely remove essential object features is crucial for accurate detection or segmentation, (2) Realistic Occlusions: Creating occlusion patterns that mimic real-world scenarios can be challenging. (3) Data Imbalance: Over-occluding objects might lead to imbalanced training data, (4) Computational Efficiency: Generating complex occlusions can be computationally expensive. Some strategies for addressing challenges are (1) Occlusion Probability [11]: Controlling the probability of occlusions can help balance the dataset, (2) Occlusion Size and Shape [12]: Varying the size and shape of occluded regions can increase diversity, (3) Occlusion Content [13]: Using meaningful objects or textures for occlusions can improve realism, (4) Combination with Other Augmentations [14]: Combining occlusion with other augmentations can enhance model robustness. By carefully considering these factors, practitioners can effectively incorporate occlusion handling into their image augmentation pipelines.

Boundary refinement is a critical aspect of image segmentation, focusing on improving the accuracy of object boundaries [15]. Augmentation techniques can significantly contribute to this process. Subtle rotations, scaling, and shearing can help the model learn to accurately represent object boundaries. Simulating real-world deformations can improve the model's ability to handle complex object shapes and boundaries. Introducing color variations can help the model distinguish objects from their backgrounds based on color cues. Adding noise to images can force the model to focus on object boundaries to accurately segment objects. For tasks like organ segmentation, applying elastic deformations can help refine organ boundaries and capture intricate details [16]. Using geometric transformations and noise addition can improve the accuracy of building or road segmentation by enhancing boundary delineation. By focusing on boundary refinement through augmentation, segmentation models can achieve more precise and accurate results, especially in applications where precise object localization is crucial. The following example demonstrates a basic implementation of elastic deformation for image augmentation. We can further refine this by incorporating more sophisticated deformation techniques and exploring different parameter combinations to optimize the augmentation process for our specific segmentation task.

```python
import numpy as np
from scipy.ndimage import gaussian_filter
def elastic_transform(image, alpha, sigma):
    """
    Applies elastic deformation to an image.
    Args:
      image: The input image (numpy array).
      alpha: The strength of the deformation.
      sigma: The standard deviation of the Gaussian kernel used for
generating the displacement field.
    Returns:
      The elastically deformed image.
    """
    shape = image.shape
    dx = gaussian_filter((np.random.rand(*shape) * 2 - 1), sigma,
mode="constant", cval=0) * alpha
    dy = gaussian_filter((np.random.rand(*shape) * 2 - 1), sigma,
mode="constant", cval=0) * alpha
    x, y = np.meshgrid(np.arange(shape[1]), np.arange(shape[0]))
    indices = np.reshape(y + dy, (-1, 1)), np.reshape(x + dx, (-1, 1))
    indices = indices.astype(np.float32)
    indices = np.clip(indices, 0, np.array(shape[0:2]) - 1)
    indices = indices.astype(np.int32)
    deformed_image = image[indices[0, :], indices[1, :]]
    deformed_image = deformed_image.reshape(image.shape)
    return deformed_image

# Example usage:
# Assuming 'image' is the input image (e.g., a numpy array)
deformed_image = elastic_transform(image, alpha=20, sigma=4)

# Apply this augmentation during training along with other relevant
augmentations
# such as rotations, scaling, and noise addition.
```

This code snippet defines a function `elastic_transform` that applies elastic deformation to an input image. It creates two displacement fields (`dx` and `dy`) using Gaussian filters. These fields represent the amount of displacement to be applied to each pixel in the x and y directions. It creates a grid of indices representing the original pixel coordinates. It adds the displacement fields (`dx` and `dy`) to the grid of indices to obtain the new coordinates for each pixel. The indices are clipped to ensure they remain within the image boundaries. The image is deformed by mapping the original pixel values to their new locations based on the calculated indices. The function returns the elastically deformed image. This function can be integrated into our image augmentation pipeline during training to improve the model's ability to handle subtle deformations and refine object boundaries. We have to adjust the `alpha` and `sigma` parameters to control the strength and smoothness of the deformation.

While augmentation can significantly improve boundary refinement, several challenges must be addressed: (1) Preserving Ground Truth [17]: Ensuring that augmentations accurately reflect changes in object boundaries in the ground truth masks is crucial, (2) Data Imbalance [18]: Over-augmenting specific boundary types can lead to imbalanced training data, (3) Computational Cost [19]: Complex augmentations like elastic deformations can be computationally expensive, especially for large image datasets, (4) Augmentation Strength [20]: Determining the optimal level of augmentation to improve image boundary refinement without introducing noise is challenging, (5) Evaluation Metrics: Evaluating the impact of augmentation on boundary accuracy requires specific metrics beyond standard segmentation metrics. Addressing these challenges requires careful experimentation and fine-tuning of augmentation parameters.

4.3 Deep Learning Frameworks and Tools for Image Augmentation

Several deep learning frameworks offer built-in image augmentation capabilities, simplifying the process for developers and researchers. TensorFlow/Keras provides a wide range of augmentation functions, including geometric transformations (rotation, flipping, cropping), color adjustments (brightness, contrast, saturation), and noise addition [21]. Keras, a high-level API built on TensorFlow, offers an intuitive interface for applying augmentations. TensorFlow/Keras can be combined with other image-processing libraries for more advanced augmentations. PyTorch offers a high degree of flexibility in creating custom augmentation transformations [22]. Torchvision includes built-in augmentation functions for common transformations. PyTorch is widely used in the research community, with a strong ecosystem of tools and libraries. Albumentations is a dedicated library Specifically designed for image augmentation, offering a comprehensive set of transformations [23]. It is optimized for performance and ease of use. This allows for creating custom augmentation pipelines. By using these frameworks and tools, developers can efficiently implement image augmentation pipelines and enhance the performance of their deep learning models.

Selecting the appropriate image augmentation framework is crucial for efficient and effective project execution. Several factors should be considered. Some of them are as follows.

Augmentation Types: Determining the appropriate augmentation techniques is crucial for optimizing model performance. The specific transformations required depend on the nature of the data, the desired outcome, and the target application [24]. Rotation, flipping, scaling, shearing, cropping, and padding can introduce variations in object orientation, size, and position. Adjustments to brightness, contrast, hue, saturation, and color channels simulate different lighting conditions. Adding Gaussian, salt-and-pepper, or other types of noise can improve model robustness. Simulating real-world deformations to enhance model generalization. Removing random image patches to increase model robustness. By carefully selecting augmentation types and considering their impact on the target application, practitioners can significantly enhance model performance.

Dataset Size: The ability to handle large datasets efficiently is a critical factor in selecting an image augmentation framework. With limited data, simple transformations like flips, rotations, and crops can be sufficient to increase data diversity. For limited datasets, over-aggressive augmentation can lead to data degradation and hinder model performance [25]. Many real-world applications involve massive amounts of image data, requiring frameworks capable of processing terabytes of information. Support for parallel processing and distributed computing can accelerate augmentation on large datasets. Efficient handling of image batches is essential for maximizing hardware utilization and reducing processing time. Frameworks should effectively manage memory usage to avoid out-of-memory errors, especially when dealing with high-resolution images. By understanding the relationship between dataset size and augmentation techniques, practitioners can effectively use augmentation to improve model performance. For example, when working with a dataset of millions of images, a framework that can process images in batches of thousands, while effectively utilizing GPU memory, would be ideal.

Performance: Evaluating the speed and efficiency of an image augmentation framework is crucial for selecting the optimal tool for a project [23]. The framework should be able to apply augmentations to large datasets in a reasonable amount of time. The framework should effectively utilize available computational resources, such as GPUs or TPUs. The framework should minimize memory usage to avoid bottlenecks. Benchmarking different frameworks with representative datasets can help identify the most performant options. Key metrics to evaluate are: (1) Augmentation time per image which measures the speed of applying augmentations to individual images, (2) Throughput which evaluates the number of images processed per second, (3) Memory usage which assesses the amount of memory required for augmentation. By carefully considering these performance metrics, practitioners can select a framework that aligns with their project's specific requirements and computational resources.

Flexibility: A crucial aspect of an image augmentation framework is its flexibility to accommodate custom transformations [26]. This capability is essential for researchers and practitioners who require tailored augmentations for specific domains or applications. The framework should allow users to fine-tune augmentation parameters to achieve desired effects. The ability to define and integrate custom augmentation logic is crucial for complex scenarios. The framework should support combining multiple augmentations in various sequences to create complex pipelines. The framework should be open to incorporating new augmentation techniques as they are developed. By offering these features, a flexible framework empowers users to experiment with different augmentation strategies and optimize performance for their specific tasks.

Built-in Augmentations: Deep learning frameworks often provide a rich set of pre-defined augmentation functions, streamlining the process of data enrichment. These built-in augmentations cover a wide range of transformations, including geometric manipulations (rotation, flipping, cropping, shearing), color adjustments (brightness, contrast, hue, saturation), and noise injection (Gaussian, salt-and-pepper). By using these pre-implemented functions, developers can rapidly experiment with different augmentation strategies without the need for extensive custom implementation. However, it's vital to consider the specific requirements of the project and explore additional augmentation techniques if necessary. Here are some examples [21, 22]:

TensorFlow/Keras.

- tf.image.random_flip_left_right
- tf.image.random_crop
- tf.image.random_brightness
- tf.image.random_contrast
- tf.image.random_hue
- tf.image.random_saturation

PyTorch

- torchvision.transforms.RandomHorizontalFlip
- torchvision.transforms.RandomCrop
- torchvision.transforms.RandomRotation
- torchvision.transforms.ColorJitter
- torchvision.transforms.RandomAffine

These functions provide a solid foundation for image augmentation, allowing developers to quickly experiment with different transformations and optimize their models. While these frameworks offer a good starting point, custom augmentation functions might be necessary for specific use cases or to achieve desired levels of complexity.

Custom Augmentations: While pre-built augmentation functions offer a solid foundation, many image augmentation tasks demand tailored solutions. This is where the flexibility to create custom augmentations becomes crucial. Some applications require specialized augmentations that align with the problem domain. For instance, in medical imaging, simulating specific types of noise or deformations might be necessary. Combining multiple augmentation steps or applying complex transformations can create highly diverse and realistic augmented images. Custom augmentations allow researchers to explore new augmentation strategies and optimize performance. It is possible by direct code implementation using the underlying image-processing libraries (e.g., OpenCV, PIL) to create custom augmentation functions [27]. For example, to create a custom image rotation function, first import OpenCV and NumPy for image-processing and numerical operations. Then we have to create a function "random_rotation" that takes an image and an optional angle range as input and generates a random angle within the specified range. Next, we can use OpenCV's "cv2.getRotationMatrix2D" to create a rotation matrix. Then we can use

OpenCV's "cv2.warpAffine" to apply the rotation to the image. At last, Return the rotated image as the output. The corresponding Python code is as follows.

```python
import cv2
import numpy as np
def random_rotation(image, angle_range=(-20, 20)):
    """
    Rotates an image by a random angle within the specified range.
    Args:
      image: The input image as a NumPy array.
      angle_range: A tuple specifying the minimum and maximum rotation
angle.
    Returns:
      The rotated image as a NumPy array.
    """
    height, width = image.shape[:2]
    angle = np.random.randint(angle_range[0], angle_range[1])
    M = cv2.getRotationMatrix2D((width / 2, height / 2), angle, 1)
    rotated_image = cv2.warpAffine(image, M, (width, height))
    return rotated_image
```

Many deep learning frameworks offer hooks or extension points to incorporate custom augmentation functions. This provides flexibility and allows developers to integrate complex or domain-specific augmentations into their pipelines. Albumentations, a popular image augmentation library, provides a Compose class that allows users to create custom augmentation pipelines. This class can be used to combine built-in and custom augmentations seamlessly. Consider the following code:

```python
import albumentations as A
# Define a custom augmentation function
def my_custom_augmentation(image, **params):
    # Apply custom augmentation logic here
    return augmented_image
# Create a custom augmentation
custom_transform = A.Lambda(image=my_custom_augmentation)
# Compose multiple augmentations
transform = A.Compose([
    custom_transform,
    A.RandomCrop(width=224, height=224),
    A.HorizontalFlip(p=0.5)
])
# Apply the transformation to an image
augmented_image = transform(image=image)
```

In this example, a custom augmentation function my_custom_augmentation is defined, which can be incorporated into the augmentation pipeline using the A.Lambda transform. This approach offers flexibility to experiment with different augmentation strategies and integrate domain-specific knowledge. By harnessing the power of custom augmentations, researchers can push the boundaries of image augmentation and achieve superior model performance.

Batch Processing: Batch processing is essential for handling large datasets efficiently when performing image augmentation [28]. Deep learning frameworks and libraries should be capable of processing multiple images simultaneously to accelerate training and reduce computational overhead. Efficient batch processing involves optimizing data loading, augmentation pipeline execution, and memory management. Frameworks like TensorFlow and PyTorch offer built-in support for batch processing, allowing users to process large datasets in smaller, manageable chunks. By effectively utilizing batch processing, practitioners can significantly improve the training speed and overall efficiency of their image augmentation pipelines. Consider the following example of batch processing:

```python
import torch
import torchvision.transforms as transforms
# Define image transformations
transform = transforms.Compose([
    transforms.RandomCrop(32, padding=4),
    transforms.RandomHorizontalFlip(),
    transforms.ToTensor(),
    transforms.Normalize((0.5, 0.5, 0.5), (0.5, 0.5, 0.5))
])

# Load data into a dataset
dataset = torchvision.datasets.CIFAR10(root='./data', train=True,
download=True, transform=transform)
# Create a data loader for batch processing
dataloader = torch.utils.data.DataLoader(dataset, batch_size=64,
shuffle=True, num_workers=2)
# Iterate through batches
for images, labels in dataloader:
    # Process the batch of images
    # ...
```

In this example, a batch size of 64 is used, meaning 64 images are processed at once. This improves efficiency compared to processing images individually. By effectively utilizing batch processing, practitioners can significantly accelerate image augmentation pipelines and optimize training processes.

Integration: Seamless integration of image augmentation into the deep learning pipeline is crucial for efficient model training and evaluation. Modern frameworks like TensorFlow, PyTorch, and Keras provide robust tools for this purpose. These frameworks offer data loaders that can handle image augmentation on the fly, reducing memory overhead and improving training efficiency. Most frameworks allow users to define custom augmentation pipelines, providing flexibility in tailoring augmentations to specific tasks. Augmentation can be easily incorporated into training scripts, ensuring consistency between data preprocessing and model training. Many frameworks support GPU acceleration for image augmentation, significantly improving performance. Consider the following example of integrating image augmentation directly into the training pipeline using TensorFlow and Keras.

```
from tensorflow.keras.preprocessing.image import ImageDataGenerator
train_datagen = ImageDataGenerator(
    rescale=1./255,
    rotation_range=40,
    width_shift_range=0.2,
    height_shift_range=0.2,
    shear_range=0.2,
    zoom_range=0.2,
    horizontal_flip=True,
    fill_mode='nearest')
train_generator = train_datagen.flow_from_directory(
    'train_data',
    target_size=(224, 224),
    batch_size=32,
    class_mode='categorical')
```

This class allows for real-time data augmentation during training. It supports a wide range of augmentations, including rotation, flipping, zooming, shearing, and brightness/contrast adjustments.

PyTorch provides similar functionalities through the `torchvision.transforms` module. Here is an example of this.

```
import torchvision.transforms as transforms
transform = transforms.Compose([
    transforms.RandomCrop(32, padding=4),
    transforms.RandomHorizontalFlip(),
    transforms.ToTensor(),
    transforms.Normalize((0.5, 0.5, 0.5), (0.5, 0.5, 0.5)),
])
trainset = torchvision.datasets.CIFAR10(root='./data', train=True,
download=True, transform=transform)
trainloader = torch.utils.data.DataLoader(trainset, batch_size=4,
shuffle=True, num_workers=2)
```

By effectively integrating image augmentation into the deep learning pipeline, practitioners can significantly enhance model performance and robustness.

Active Community: A vibrant and active community surrounding image augmentation tools and techniques is invaluable for researchers and practitioners [29]. Such communities offer a wealth of resources, including code snippets, tutorials, and best practices. They serve as platforms for knowledge sharing and troubleshooting, allowing users to learn from the experiences of others. Additionally, active communities often contribute to the development of new augmentation techniques and libraries, ensuring that the field continues to advance. By participating in these communities, practitioners can access valuable insights, accelerate development, and stay updated on the latest trends in image augmentation. For instance, platforms like Kaggle and GitHub host numerous datasets, code repositories, and discussions related to image augmentation. These communities facilitate the sharing of best practices, troubleshooting, and benchmarking. Additionally, academic conferences and workshops provide opportunities for researchers to present their work and engage in discussions with peers.

Documentation: Poor documentation of image augmentation tools and techniques can significantly hinder project progress and lead to suboptimal results. Without clear explanations and examples, developers spend more time understanding and experimenting with different augmentation options, slowing down the project timeline. Lack of documentation can lead to inconsistent application of augmentation techniques, resulting in unpredictable outcomes and difficulties in reproducing results. Misunderstanding augmentation parameters or applying incorrect transformations can introduce errors into the dataset and affect model performance. Poor documentation might restrict users to basic augmentations, preventing exploration of more advanced techniques. Without proper documentation, it's challenging to share knowledge and best practices within teams or with the broader community. Thus, comprehensive documentation is essential for efficient and effective image augmentation, enabling researchers and practitioners to maximize the potential of these techniques. Clear explanations of available functions, parameters, and expected outputs enable users to know the influence of different augmentations on their data. The key aspects of good documentation are (1) Clear and concise explanations of each augmentation technique, including mathematical formulas or visualizations where applicable, (2) Recommendations on how to adjust augmentation parameters based on dataset characteristics and desired outcomes, (3) Code snippets and illustrative examples demonstrating how to apply augmentations effectively, (4) Information about the computational cost and potential impact of augmentations on training time, (5) Guidelines on combining different augmentations for optimal results. By providing thorough documentation, tool developers empower users to make informed decisions and experiment with various augmentation strategies.

Updates and Maintenance: The dynamic nature of deep learning necessitates continuous updates and improvements in image augmentation frameworks [30]. Active maintenance ensures that these tools remain relevant and effective. Regular updates introduce new augmentation techniques, optimize performance, and address compatibility issues with emerging deep learning frameworks. A thriving development community contributes to the evolution of these tools, incorporating user feedback and incorporating state-of-the-art advancements. By staying updated with the latest developments, practitioners can benefit from new features, bug fixes, and performance enhancements, ultimately improving the quality and efficiency of their image augmentation pipelines. For example, Albumentations exemplifies the importance of continuous updates and maintenance. The library has evolved from providing basic geometric transformations to incorporating advanced techniques like cutout, random erasing, and grid distortion. Regular updates address performance optimizations, compatibility issues with newer deep learning frameworks, and user feedback. For instance, the introduction of new augmentation techniques like grid distortion and mosaic augmentation demonstrates the library's commitment to staying at the forefront of image augmentation research. By actively maintaining and updating the library, Albumentations confirms that users have access to the newest advancements in image augmentation. This ongoing development fosters a thriving community of users who contribute to the library's growth and share their experiences, creating a positive feedback loop that benefits all users.

By carefully evaluating these factors and considering the specific needs of our project, we can select the most suitable image augmentation framework.

4.4 Summary

In this chapter, I discussed different applications of deep learning-based image augmentation. Deep learning-based image augmentation significantly enhances the performance of various computer vision tasks. By artificially expanding training datasets, augmentation improves model generalization, robustness, and accuracy. Augmentation addresses data scarcity, improves model robustness to variations in lighting, scale, and orientation, and enhances feature learning. Augmentations like scaling, rotation, cropping, and occlusion simulation improve object localization, handle object variations, and enhance model robustness. Augmentation techniques like geometric transformations, color adjustments, and noise addition refine object boundaries, handle imbalanced data, and improve segmentation accuracy. By effectively applying image augmentation techniques, practitioners can develop more accurate, reliable, and robust deep learning models for a wide range of applications.

References

1. Kumar, T., Brennan, R., Mileo, A., & Bendechache, M. (2024). Image data augmentation approaches: A comprehensive survey and future directions. *IEEE Access*.
2. Shorten, C., & Khoshgoftaar, T. M. (2019). A survey on image data augmentation for deep learning. *Journal of Big Data, 6*(1), 1–48.
3. Archana, R., & Jeevaraj, P. E. (2024). Deep learning models for digital image processing: A review. *Artificial Intelligence Review, 57*(1), 11.
4. Islam, T., Hafiz, M. S., Jim, J. R., Kabir, M. M., & Mridha, M. F. (2024). A systematic review of deep learning data augmentation in medical imaging: Recent advances and future research directions. *Healthcare Analytics*, 100340.
5. Akhtar, N., & Mian, A. (2018). Threat of adversarial attacks on deep learning in computer vision: A survey. *Ieee Access, 6*, 14410–14430.
6. Zhao, Z. Q., Zheng, P., Xu, S. T., & Wu, X. (2019). Object detection with deep learning: A review. *IEEE transactions on neural networks and learning systems, 30*(11), 3212–3232.
7. Zhang, J., Wu, J., Zhou, X. S., Shi, F., & Shen, D. (2023, September). Recent advancements in artificial intelligence for breast cancer: Image augmentation, segmentation, diagnosis, and prognosis approaches. In *Seminars in Cancer Biology*. Academic Press.
8. Li, K., Yu, L., & Heng, P. A. (2022). Towards reliable cardiac image segmentation: Assessing image-level and pixel-level segmentation quality via self-reflective references. *Medical Image Analysis, 78*, 102426.
9. Saleh, K., Szénási, S., & Vámossy, Z. (2021, January). Occlusion handling in generic object detection: A review. In *2021 IEEE 19th World Symposium on Applied Machine Intelligence and Informatics (SAMI)* (pp. 000477–000484). IEEE.

10. Ouardirhi, Z., Mahmoudi, S. A., & Zbakh, M. (2024). Enhancing object detection in smart video surveillance: A survey of occlusion-handling approaches. *Electronics, 13*(3), 541.
11. Ibrahim, M., Rautek, P., Reina, G., Agus, M., & Hadwiger, M. (2021). Probabilistic occlusion culling using confidence maps for high-quality rendering of large particle data. *IEEE Transactions on Visualization and Computer Graphics, 28*(1), 573–582.
12. Zhang, T., Huang, B., & Wang, Y. (2020). Object-occluded human shape and pose estimation from a single color image. In *Proceedings of the IEEE/CVF conference on computer vision and pattern recognition* (pp. 7376–7385).
13. Macedo, M. C., & Apolinario, A. L. (2021). Occlusion handling in augmented reality: Past, present and future. *IEEE Transactions on Visualization and Computer Graphics, 29*(2), 1590–1609.
14. Rebuffi, S. A., Gowal, S., Calian, D. A., Stimberg, F., Wiles, O., & Mann, T. A. (2021). Data augmentation can improve robustness. *Advances in Neural Information Processing Systems, 34*, 29935–29948.
15. Marmanis, D., Schindler, K., Wegner, J. D., Galliani, S., Datcu, M., & Stilla, U. (2018). Classification with an edge: Improving semantic image segmentation with boundary detection. *ISPRS Journal of Photogrammetry and Remote Sensing, 135*, 158–172.
16. Han, Z., & Dou, Q. (2024). A review on organ deformation modeling approaches for reliable surgical navigation using augmented reality. *Computer Assisted Surgery, 29*(1), 2357164.
17. Bosquet, B., Cores, D., Seidenari, L., Brea, V. M., Mucientes, M., & Del Bimbo, A. (2023). A full data augmentation pipeline for small object detection based on generative adversarial networks. *Pattern Recognition, 133*, 108998.
18. Xu, H., Yan, Z. H., Ji, B. W., Huang, P. F., Cheng, J. P., & Wu, X. D. (2022). Defect detection in welding radiographic images based on semantic segmentation methods. *Measurement, 188*, 110569.
19. Chlap, P., Min, H., Vandenberg, N., Dowling, J., Holloway, L., & Haworth, A. (2021). A review of medical image data augmentation techniques for deep learning applications. *Journal of Medical Imaging and Radiation Oncology, 65*(5), 545–563.
20. Lalitha, V., & Latha, B. (2022). A review on remote sensing imagery augmentation using deep learning. *Materials Today: Proceedings, 62*, 4772–4778.
21. Shanmugamani, R. (2018). *Deep learning for computer vision: Expert techniques to train advanced neural networks using TensorFlow and Keras.* Packt Publishing Ltd.
22. Stevens, E., Antiga, L., & Viehmann, T. (2020). *Deep learning with PyTorch.* Manning Publications.
23. Buslaev, A., Iglovikov, V. I., Khvedchenya, E., Parinov, A., Druzhinin, M., & Kalinin, A. A. (2020). Albumentations: Fast and flexible image augmentations. *Information, 11*(2), 125.
24. Yue, T., Briand, L. C., & Labiche, Y. (2011). A systematic review of transformation approaches between user requirements and analysis models. *Requirements Engineering, 16*, 75–99.
25. Gul, S., Asif, M., Saleem, K., & Imran, M. (2025). Advancing aspect-based sentiment analysis in course evaluation: A multi-task learning framework with selective paraphrasing. *IEEE Access.*
26. Garcea, F., Serra, A., Lamberti, F., & Morra, L. (2023). Data augmentation for medical imaging: A systematic literature review. *Computers in Biology and Medicine, 152*, 106391.
27. Villán, A. F. (2019). *Mastering OpenCV 4 with Python: A practical guide covering topics from image processing, augmented reality to deep learning with OpenCV 4 and Python 3.7.* Packt Publishing Ltd.
28. Maharana, K., Mondal, S., & Nemade, B. (2022). A review: Data pre-processing and data augmentation techniques. *Global Transitions Proceedings, 3*(1), 91–99.
29. Booth, S. E., & Kellogg, S. B. (2015). Value creation in online communities for educators. *British Journal of Educational Technology, 46*(4), 684–698.
30. Alom, M. Z., et al. (2019). A state-of-the-art survey on deep learning theory and architectures. *electronics, 8*(3), 292.

Chapter 5
Evaluating and Optimizing Deep Learning Image Augmentation Strategies

Data augmentation has emerged as a cornerstone of successful deep learning models in computer vision, significantly enhancing their robustness, generalization, and overall performance. While various augmentation techniques have proven effective, the key to unlocking their full potential lies in careful evaluation and optimization. This chapter will delve into crucial aspects of evaluating and optimizing deep learning image augmentation strategies, exploring key considerations, metrics, and best practices for selecting and combining augmentation methods to achieve optimal model performance and improve the overall robustness and generalizability of deep learning models.

5.1 Evaluation of Image Augmentation Techniques

Evaluating the effectiveness of image augmentation techniques is crucial for optimizing model performance. Several metrics and approaches can be employed to assess augmentation impact. Some of them are discussed here.

5.1.1 Downstream Task Metrics

Evaluating the impact of image augmentation on model performance requires careful consideration of appropriate metrics. These metrics vary depending on the specific task. The following are specifically used for image classification.

A **confusion matrix (CM)** [1] is a performance evaluation tool for classification models. It provides a detailed breakdown of correct and incorrect predictions across different classes. A confusion matrix is typically a square matrix where rows represent the actual classes and columns represent the predicted classes. Elements

© The Author(s), under exclusive license to Springer Nature Singapore Pte Ltd. 2025
J. Chaki, *The Art of Deep Learning Image Augmentation: The Seeds of Success*,
SpringerBriefs in Computational Intelligence,
https://doi.org/10.1007/978-981-96-5081-1_5

of a Confusion Matrix are (1) True Positive (TP): Correctly predicted positive class instances, (2) True Negative (TN): Correctly predicted negative class instances, (3) False Positive (FP): Incorrectly predicted as positive (Type I error) and (4) False Negative (FN): Incorrectly predicted as negative (Type II error). A confusion matrix provides a comprehensive overview of model performance, enabling analysis beyond simple accuracy metrics. By using the CM, we can identify the class-wise performance and imbalances. Also, we can analyze error patterns to improve model performance. By utilizing confusion matrices, practitioners can gain valuable insights into the strengths and weaknesses of their classification models.

Accuracy [2] is a straightforward metric to evaluate the performance of an image classification model. It measures the proportion of correctly classified images out of the total number of images.

$$\text{Accuracy} = \frac{\text{Number of Correct Predictions}}{\text{Total Number of Predictions}} \tag{5.1}$$

While accuracy is a simple metric to calculate, it can be misleading in imbalanced datasets. For example, if a dataset contains 90% negative samples and 10% positive samples, a model that always predicts the majority class (negative) can achieve high accuracy without being truly effective. Therefore, it's essential to use accuracy in conjunction with other metrics like precision, recall, and F1-score for a more comprehensive evaluation of model performance.

Precision [3] measures the accuracy of positive predictions made by a classification model. It answers the question: "Of all the instances predicted as positive, how many were positive?"

$$\text{Precision} = \frac{\text{True Positives}}{\text{True Positives} + \text{False Positives}} \tag{5.2}$$

Consider a model that classifies images as either "cat" or "dog." If the model predicts 10 images as cats, and out of those 10, only 7 are cats, then the precision would be: $7 / (7 + 3) = 0.7$. In this case, the model correctly identified 7 out of 10 predicted cat images, resulting in a precision of 0.7 or 70%. A high precision indicates that when the model predicts an image as a cat, it is likely to be correct. Precision is often used in conjunction with recall to evaluate model performance comprehensively. A high precision might indicate a conservative model that is less likely to make false positive errors. Understanding precision is essential for assessing the reliability of positive predictions made by a classification model.

Recall [4], also known as sensitivity, measures a model's ability to correctly identify all positive instances. It answers the question: "Of all the actual positive cases, how many did the model correctly identify?"

$$\text{Recall} = \frac{\text{True Positives}}{\text{True Positives} + \text{False Negatives}} \tag{5.3}$$

In a medical image classification task for detecting tumors, recall measures the model's ability to find all actual tumors in the dataset. A high recall indicates that the model is good at identifying all tumor cases and minimizing false negatives. A high recall is crucial in applications where missing positive instances is costly, such as in medical diagnosis. Recall is often used in conjunction with precision to evaluate model performance comprehensively. Understanding recall is essential for assessing a model's ability to capture all relevant instances within a dataset.

The **F1-score** [5] provides a single metric that combines the information from both precision and recall. It is particularly useful when there is an imbalance between positive and negative classes.

$$\text{F1 Score} = 2 \times \frac{(\text{Precision} \times \text{Recall})}{(\text{Precision} + \text{Recall})} \tag{5.4}$$

In a medical image classification task for detecting lung cancer, a high F1-score indicates that the model effectively balances the ability to correctly identify cancer cases (precision) and avoid missing cancer cases (recall). The F1-score is a harmonic mean, giving more weight to lower values. This means that for an F1 score to be high, both precision and recall must be reasonably high. An F1 score of 1 represents perfect precision and recall, while a score of 0 indicates either perfect precision with zero recall or vice versa. By using the F1-score, we can get a more comprehensive understanding of the model's performance compared to using precision or recall alone.

The following evaluations are specifically used for object detection. **Mean Average Precision (mAP)** [6] is a comprehensive metric used to evaluate object detection models. It considers both object localization and classification accuracy. The average precision (AP) for a single class is calculated by computing the precision-recall curve and interpolating the precision values at recall points spaced equally (e.g., 0.1, 0.2, …, 1). The AP is then the average of these interpolated precision values. The mAP is the average of the AP values calculated for each class.

$$\text{mAP} = \frac{(AP_{\text{class1}} + AP_{\text{class2}} + \cdots + AP_{\text{classN}})}{N}, \tag{5.5}$$

where mAP is the mean average precision, AP_{class} is the average precision for a specific class and N is the number of classes.

Consider an object detection model for detecting cars, pedestrians, and bicycles in images. The mAP is calculated by computing the average precision for each class (cars, pedestrians, and bicycles) and then averaging these values. mAP considers both precision and recall, making it a robust metric for object detection. It is widely used in object detection competitions and benchmarks. Higher mAP values indicate better model performance. By calculating mAP, we can obtain a comprehensive evaluation of an object detection model's ability to accurately locate and classify objects in images.

Intersection over Union (IoU), [7] also known as the Jaccard Index, is a metric used to evaluate the overlap between predicted and ground truth bounding boxes in object detection. It quantifies the degree of overlap between the two boxes.

$$\text{IoU} = \frac{\text{Area of Overlap}}{\text{Area of Union}}, \tag{5.6}$$

where the Area of Overlap is the area of the region common to both bounding boxes and the Area of Union is the total area covered by both bounding boxes.

Consider two bounding boxes: a ground truth box and a predicted box. If the predicted box perfectly overlaps the ground truth box, the IoU is 1.0, indicating a perfect match. If there is no overlap, the IoU is 0. IoU is a more robust metric than simple metrics like accuracy because it considers the spatial overlap between the bounding boxes. It is commonly used in object detection tasks to evaluate the localization accuracy of the model. IoU values typically range from 0 to 1, with higher values indicating better overlap. By calculating IoU for each detected object and averaging the scores, we can obtain a comprehensive evaluation of the object detection model's localization performance.

While IoU is a valuable metric for evaluating object detection models, it has certain limitations: (1) Sensitivity to Scale: Small objects with high IoU scores might not be as significant as larger objects with lower IoU scores, (2) Insensitivity to Spatial Relationships: IoU doesn't consider the spatial relationship between predicted and ground truth boxes beyond overlap, which can lead to misleading evaluations in certain scenarios, (3) Limited Information: When IoU is zero (no overlap), it provides no information about the quality of the prediction, (4) Difficulty in Handling Multiple Objects: In cases where multiple objects are present, calculating IoU for each object and then averaging can be complex. To address these limitations, other metrics like Generalized IoU (GIoU), Distance IoU (DIoU), and Complete IoU (CIoU) have been proposed. These metrics incorporate additional information about the bounding boxes, such as their distance, aspect ratio, and center points, to provide a more comprehensive evaluation.

Generalized Intersection over Union (GIoU) [8] is an improvement over the standard IoU metric, addressing some of its limitations. It considers not only the overlap between the predicted and ground truth bounding boxes but also the relationship between their enclosing boxes.

$$\text{GIoU} = \text{IoU} - \left(\text{Area}(C) - \frac{\text{Area}(\text{Union}(A, B))}{\text{Area}(C)} \right), \tag{5.7}$$

where IoU is the Intersection over Union, A is the predicted bounding box, B is the ground truth bounding box and C is the smallest bounding box enclosing both A and B. The GIoU penalizes the model for generating bounding boxes that are far from the ground truth, even when there is some overlap. The additional term in the GIoU equation encourages the model to produce bounding boxes that are closer to the ground truth.

Consider two bounding boxes: one predicted and one ground truth. If the predicted box is completely inside the ground truth box but has a smaller size, the IoU will be high, but the GIoU will be lower due to the additional penalty term. This encourages the model to generate bounding boxes that are closer in size and position to the ground truth. This method encourages models to generate tighter bounding boxes and can be used as a loss function for object detection models. Also, GIoU addresses the limitations of IoU by considering the overall spatial relationship between bounding boxes. By incorporating GIoU into the evaluation process, we can obtain a more informative metric for object detection models.

Distance IoU (DIoU) [9] is an improvement over GIoU that explicitly considers the Euclidean distance between the centers of the ground truth and predicted bounding boxes.

$$\text{DIoU} = \text{IoU} - \frac{\rho^2(b, b_{gt})}{c^2}, \tag{5.8}$$

where IoU is the Intersection over Union, $\rho(b, b_{gt})$ is the Euclidean distance between the centers of the bounding boxes and c is the diagonal length of the smallest enclosing box covering both boxes. The DIoU loss not only considers the overlap between the boxes but also the distance between their centers. This encourages the model to predict bounding boxes that are closer to the ground truth center, leading to faster convergence and better performance, especially when the bounding boxes do not overlap.

Consider two bounding boxes that do not overlap. GIoU would only consider the overlap area, which is zero, providing no information about the relative positions of the boxes. DIoU, on the other hand, would penalize the distance between the centers of the boxes, encouraging the model to generate bounding boxes closer to the ground truth. DIoU encourages faster convergence and can be used as a loss function for object detection. By incorporating the distance between bounding box centers, DIoU provides a more informative metric for object detection compared to IoU and GIoU.

Complete IoU (CIoU) [10] is an extension of DIoU that incorporates an additional term to penalize the aspect ratio difference between the predicted and ground truth bounding boxes.

$$\text{CIoU} = \text{IoU} - \left(\frac{\rho^2(b, b_{gt})}{c^2} \right) - \alpha v, \tag{5.9}$$

where IoU is the Intersection over Union, $\rho(b, b_{gt})$ is the Euclidean distance between the centers of the bounding boxes, c is the diagonal length of the smallest enclosing box covering both boxes, α is a weight function, v is a measure of the aspect ratio difference. The CIoU loss not only considers the overlap between the boxes and the distance between their centers but also penalizes differences in aspect ratio. This encourages the model to predict bounding boxes with the correct sizes and shapes.

The following are the evaluation matrices concerning the image segmentation task. **Pixel Accuracy** [11] is a straightforward metric to evaluate the overall performance of an image segmentation model. It calculates the percentage of correctly classified pixels in an image.

$$\text{Pixel Accuracy} = \frac{\text{Number of Correctly Classified Pixels}}{\text{Total Number of Pixels}} \tag{5.10}$$

Consider a binary segmentation task where pixels are classified as either foreground or background. If an image contains 1000 pixels and 900 pixels are correctly classified, the pixel accuracy would be: Pixel Accuracy = 900/1000 = 0.9 or 90%

While pixel accuracy is simple to compute, it has limitations [12]: (1) **Insensitivity to Spatial Information**: Pixel accuracy does not consider the spatial arrangement of pixels. A model that correctly classifies isolated pixels but fails to capture object boundaries can still achieve high pixel accuracy, (2) **Vulnerability to Class Imbalance**: In images with large background regions, a model that consistently predicts the background class can achieve high pixel accuracy even with poor performance on foreground objects, (3) **Limited Insights**: Pixel accuracy provides a general overview of model performance but doesn't offer detailed information about specific errors or the nature of segmentation mistakes. To address these limitations, other metrics like Intersection over Union (IoU) and mean IoU are often used in conjunction with pixel accuracy to provide a more comprehensive evaluation of segmentation models.

Mean Intersection over Union (mIoU) [13] is a commonly used metric for evaluating image segmentation models. It calculates the average IoU across all classes in a dataset.

$$\text{mIoU} = \frac{\text{IoU}_{\text{class1}} + \text{IoU}_{\text{class2}} + \cdots + \text{IoU}_{\text{class}N}}{N}, \tag{5.11}$$

where mIoU is the mean Intersection over Union, $\text{IoU}_{\text{class}}$ is the Intersection over Union for a specific class, N is the number of classes.

Consider a segmentation task with three classes: background, car, and pedestrian. The mIoU is calculated by computing the IoU for each class and then averaging the results. A higher mIoU indicates better overall segmentation performance. The mIoU provides a comprehensive evaluation of segmentation performance across different classes and it is suitable for multi-class segmentation tasks.

While mIoU is a valuable metric for evaluating image segmentation models, it has several limitations: (1) Class Imbalance: If one class dominates the dataset, it can skew the mIoU score, making it less informative, (2) Small Objects: mIoU might not accurately reflect the performance on small objects, as even a few misclassified pixels can significantly impact the IoU score, (3) Overlapping Objects: In cases where objects overlap, calculating IoU can become ambiguous, leading to inaccurate evaluation, (4) Ignore Spatial Relationships: mIoU doesn't consider the spatial distribution of errors, which can be crucial for certain applications. To address these limitations, other metrics like Frequency Weighted IoU (FWIoU) and Panoptic Quality (PQ) have been proposed.

Frequency Weighted IoU (FWIoU) [14] is a metric designed to address the class imbalance issue inherent in mIoU. It weights the IoU of each class based on its frequency in the dataset.

$$\text{FWIoU} = \frac{\Sigma(p_i \times IoU_i)}{\Sigma p_i}, \tag{5.12}$$

where FWIoU is the Frequency Weighted IoU, p_i is the frequency of class i in the dataset and IoU_i is the IoU for class i.

Consider a segmentation task with three classes: background, car, and pedestrian. The FWIoU would calculate the IoU for each class, weight it by the frequency of that class in the dataset, and then average the weighted IoUs. FWIoU addresses class imbalance by giving more weight to frequent classes. It provides a more balanced evaluation of segmentation performance. FWIoU is an improvement over standard mIoU by incorporating class frequency information. However, it still has limitations and might not be sufficient for complex segmentation tasks.

Panoptic Quality (PQ) [15] is a metric designed to evaluate the performance of panoptic segmentation models, which combine semantic and instance segmentation. It provides a unified measure for both "stuff" (semantic) and "thing" (instance) categories.

$$PQ = \frac{\left(\sum TP \times IoU(p, g)\right)}{\left(\sum TP + 0.5 \times \sum FP + 0.5 \times \sum FN\right)}, \tag{5.13}$$

where TP is the set of true positive segments, $IoU(p, g)$ is the Intersection over Union between a predicted segment p and its matched ground truth segment g, FP is the set of false positive segments and FN is the set of false negative segments.

PQ incorporates both segmentation quality (IoU) and recognition quality (matching predicted segments to ground truth segments) into a single metric. A higher PQ indicates better overall performance in both semantic and instance segmentation. Computing PQ can be computationally expensive for large datasets. This requires careful matching of predicted and ground truth segments. PQ is a valuable metric for assessing the performance of panoptic segmentation models, providing a holistic evaluation of the model's ability to accurately segment and classify objects in a scene.

The **Dice coefficient** [16], also known as the Sørensen-Dice coefficient, is a similarity metric often used to compare the similarity of two sets of data. In the context of image segmentation, it measures the overlap between the predicted segmentation and the ground truth segmentation.

$$\text{Dice Coefficient} = \frac{(2 \times |A \cap B|)}{(|A| + |B|)}, \tag{5.14}$$

where $|A|$ is the number of pixels in the predicted segmentation, $|B|$ is the number of pixels in the ground truth segmentation, and $|A \cap B|$ is the number of pixels common to both segmentations.

Imagine a binary segmentation task where pixels are classified as either foreground or background. The Dice coefficient would measure the overlap between the predicted foreground pixels and the ground truth foreground pixels. A Dice coefficient of 1 indicates perfect overlap, while a value of 0 indicates no overlap. The Dice coefficient is closely related to the F1 score. It is sensitive to both false positives and false negatives. It is commonly used in medical image segmentation due to its ability to handle imbalanced datasets. Like IoU, the Dice coefficient can be sensitive to small objects and class imbalance. It might not capture all aspects of segmentation performance, such as spatial relationships between objects. Despite these limitations, the Dice coefficient remains a widely used metric for evaluating image segmentation models due to its simplicity and effectiveness.

By carefully selecting and applying these metrics, practitioners can assess the effectiveness of image augmentation techniques in improving model performance. The following code snippet provides a starting point for evaluating the performance of our deep learning model on augmented data. By carefully analyzing these metrics, we can gain valuable insights into the effectiveness of our augmentation strategies and identify areas for improvement.

```python
import tensorflow as tf
from tensorflow.keras.models import load_model
from sklearn.metrics import accuracy_score, precision_score,
recall_score, f1_score
def evaluate_image_augmentation(model, test_data, ground_truth_labels):
    """
    Evaluates the performance of a deep learning model on a test dataset.
    Args:
        model: The trained deep learning model.
        test_data: The test dataset containing augmented images.
        ground_truth_labels: The corresponding ground truth labels for the
test data.
    Returns:
        A dictionary containing various evaluation metrics.
    """
    predictions = model.predict(test_data)
    # Classification Metrics
    accuracy = accuracy_score(ground_truth_labels,
predictions.argmax(axis=1))
    precision = precision_score(ground_truth_labels,
predictions.argmax(axis=1), average='weighted')
    recall = recall_score(ground_truth_labels, predictions.argmax(axis=1),
average='weighted')
    f1 = f1_score(ground_truth_labels, predictions.argmax(axis=1),
average='weighted')
    # Object Detection Metrics (assuming predictions are bounding boxes)
    # ... (Implement logic for calculating IoU, GIoU, DIoU, CIoU, mAP,
mIoU, FWIoU, PQ)
    # Example: (Simplified IoU calculation)
    # Assuming predictions and ground_truth_labels are lists of bounding
boxes
    # (e.g., [x1, y1, x2, y2])
    iou_scores = []
    for pred_box, gt_box in zip(predictions, ground_truth_labels):
        # Calculate IoU for each pair of bounding boxes
        iou = calculate_iou(pred_box, gt_box)
        iou_scores.append(iou)
    mean_iou = np.mean(iou_scores)
```

```
  return {
      'accuracy': accuracy,
      'precision': precision,
      'recall': recall,
      'f1_score': f1,
      'mean_iou': mean_iou,
      # ... add other calculated metrics to the dictionary
  }
# Helper function to calculate IoU (Intersection over Union)
def calculate_iou(box1, box2):
  """
  Calculates the Intersection over Union (IoU) of two bounding boxes.
  """
  x1_min = max(box1[0], box2[0])
  y1_min = max(box1[1], box2[1])
  x2_max = min(box1[2], box2[2])
  y2_max = min(box1[3], box2[3])
  intersection_area = max(0, x2_max - x1_min) * max(0, y2_max - y1_min)
  box1_area = (box1[2] - box1[0]) * (box1[3] - box1[1])
  box2_area = (box2[2]   box2[0]) * (box2[3]   box2[1])
  union_area = box1_area + box2_area - intersection_area
  if union_area == 0:
    return 0
  else:
    return intersection_area / union_area
# Example Usage:
# 1. Load the trained model
model = load_model('path/to/our/model.h5')
# 2. Prepare the test data and ground truth labels
test_data = ...
ground_truth_labels = ...
# 3. Evaluate the model
evaluation_metrics  =  evaluate_image_augmentation(model,  test_data,
ground_truth_labels)
# 4. Print the evaluation results
print(evaluation_metrics)
```

5.1.2 Diversity Metrics

Assessing the diversity of augmented images is crucial to ensure that the augmentation process is generating a wide range of variations. Metrics like Fréchet Inception Distance (FID), Kernel Inception Distance (KID), and Shannon entropy can provide valuable insights into the effectiveness of augmentation techniques.

FID [17] is a metric used to evaluate the quality of generated images by comparing their distribution to that of real images. It measures the distance between the feature distributions of real and generated images in a high-dimensional feature space. FID is calculated based on the assumption that the feature distributions of both real and generated images can be approximated by Gaussian distributions. First features are extracted from both real and generated images using a pre-trained Inception network. Then the mean and covariance matrix of the feature vectors is calculated for both real and generated images. After that, the Fréchet distance is computed between the two Gaussian distributions.

$$\text{FID} = \left|\left|\mu_{\text{real}} - \mu_{\text{gen}}\right|\right|^2 + \text{Tr}\left(\Sigma_{\text{real}} + \Sigma_{\text{gen}} - 2\left(\Sigma_{\text{real}} \times \Sigma_{\text{gen}}\right)^{0.5}\right), \qquad (5.15)$$

where μ_{real} and μ_{gen} are the means of the feature distributions for real and generated images, respectively, Σ_{real} and Σ_{gen} are the covariance matrices of the feature distributions for real and generated images, respectively and Tr is the trace of a matrix. A lower FID value indicates a higher similarity between the real and generated image distributions, implying better image quality and diversity.

While FID is a valuable metric for evaluating models, it has several limitations: (1) FID relies on the features extracted by the Inception network, which might not capture all relevant image information, (2) FID can be sensitive to changes in the data distribution, such as image size or resolution, (3) Calculating FID can be computationally expensive for large datasets. Despite these challenges, FID remains a widely used metric for evaluating image generation models due to its simplicity and effectiveness.

KID [18] is a non-parametric alternative to FID, offering a more robust measure of the similarity between real and generated image distributions. Unlike FID, which relies on Gaussian assumptions, KID employs kernel methods to estimate the distance between distributions. KID is based on the Kernel Maximum Mean Discrepancy (MMD), which measures the distance between two probability distributions using kernel functions. The KID score is computed by calculating the MMD between the feature distributions of real and generated images extracted from a pre-trained Inception network. KID does not make assumptions about the underlying data distribution, making it more flexible. KID is often more robust to complex image distributions compared to FID. KID can detect mode collapse in generative models more effectively than FID.

KID can be used to evaluate the quality of image augmentation techniques. By comparing the KID scores of augmented images to those of the original images, we can assess the impact of augmentation on image diversity. A lower KID score indicates that the augmented images are closer in distribution to the original images, suggesting that the augmentation process preserves image characteristics.

While KID is a valuable metric for evaluating models, it has several limitations: (1) Calculating KID can be computationally expensive, especially for large datasets, (2) The choice of kernel function can influence the results, (3) Interpreting KID scores can be challenging compared to FID, which has a more intuitive interpretation.

Shannon entropy [19] is a measure of uncertainty or randomness in a probability distribution. In the context of image augmentation, it can be applied to quantify the diversity of generated images.

$$H(X) = -\Sigma p(x) * \log2(p(x)), \tag{5.16}$$

where $H(X)$ is the Shannon entropy of the random variable X and $p(x)$ is the probability of the value x.

To apply Shannon entropy to image diversity, we can extract features from images using a pre-trained network. Then discretize the feature space into bins to create a probability distribution. Lastly, the Shannon entropy is computed for the distribution of feature vectors. A higher entropy value indicates greater diversity in the generated images. The choice of feature extraction method can significantly impact the results. The number of bins used for quantization can affect the entropy calculation. Calculating entropy for large datasets can be computationally expensive. While Shannon entropy provides a theoretical foundation for measuring image diversity, its practical application in image augmentation requires careful consideration of feature representation and computational efficiency.

5.1.3 Distortion Metrics

Distortion metrics are employed to assess the degree of image degradation introduced by augmentation techniques. They help prevent excessive alterations that might hinder model performance.

Peak Signal-to-Noise Ratio (PSNR) [20] is a metric used to measure the quality of reconstructed images compared to the original images. It is widely used in image compression, image denoising, and image transmission.

$$PSNR = 20 * \log10\left(\frac{MAX_I}{RMSE}\right), \tag{5.17}$$

where PSNR is the peak signal-to-noise ratio in decibels (dB), MAX_I is the maximum pixel value of the image (e.g., 255 for 8-bit images), RMSE is the root mean squared error between the original and reconstructed images.

PSNR doesn't always correlate well with human perception of image quality. PSNR might not accurately reflect the impact of different noise types on image quality. PSNR is more sensitive to changes in bright regions of an image. Despite these limitations, PSNR remains a widely used metric due to its simplicity and computational efficiency.

Structural Similarity Index (SSIM) [21] is a perceptual image quality metric that focuses on the structural information of an image. Unlike PSNR, which is based on pixel-wise differences, SSIM considers image degradation as perceived changes in structural information. The SSIM index is calculated as the product of three components: luminance, contrast, and structure. While the exact formula is complex, it incorporates these key elements: (1) Luminance comparison which measures the similarity in brightness between the two images, (2) Contrast comparison which evaluates the difference in contrast between the two images and Structure comparison which assesses the similarity in image structure.

SSIM is often used to compare compressed images with their original versions. A high SSIM value indicates that the compressed image preserves the structural information of the original image well. SSIM correlates better with human perception of image quality than PSNR. SSIM captures changes in image structure, which is crucial for image quality assessment.

Feature Similarity Index (FSIM) [22] is a perceptual image quality assessment metric that focuses on capturing image structural information. It combines phase congruency and gradient magnitude to measure the similarity between two images. Phase congruency measures the strength of local image structures, such as edges and corners while the gradient magnitude quantifies the intensity of image gradients. FSIM calculates similarity scores based on these components and combines them to produce a final image quality measure. FSIM aligns more closely with human perception of image quality than metrics like PSNR. By emphasizing phase congruency, FSIM captures essential image information. FSIM is sensitive to various image distortions, including noise, blur, and compression artifacts. The choice of parameters for phase congruency calculation can influence the results. Also, calculating FSIM can be computationally intensive due to the involvement of phase congruency analysis. FSIM provides a valuable tool for assessing image quality, especially when structural information is critical.

Information Fidelity (IF) [23] is a metric that assesses the amount of information preserved in a distorted image relative to the original image. It quantifies the loss of information due to image processing operations, such as compression, noise reduction, or enhancement. The core idea behind IF is to measure the similarity between the feature spaces of the original and distorted images. This can be approached using various techniques, including: (1) Statistical measures: Comparing histograms, moments, or other statistical properties of the image, (2) Feature-based methods: Extracting features using techniques like SIFT, SURF, or deep learning, and then comparing the feature distributions. Consider an augmented image. A high IF value would indicate that the augmented image retains most of the information from the original image, while a low IF value suggests significant information loss. Several challenges of IF include: (1) Quantifying information content in images is challenging due to the subjective nature of perception, (2) Some methods for calculating IF can be computationally expensive, (3) The effectiveness of IF can vary depending on the type of image and the nature of the distortion. Despite these challenges, IF provides a conceptual framework for assessing image quality and information preservation.

Perceptual Image Quality Evaluator (PIQE) [24] is a deep learning-based model designed to predict human image quality assessments. It aims to bridge the gap between subjective human perception and objective image quality metrics. PIQE models are trained on a large dataset of images paired with human quality ratings. The model learns to correlate image features with human perception of quality. Given a new image, the trained PIQE model predicts a quality score based on its learned representation. It can be applied to various image degradation types. PIQE is a no-reference method, meaning it doesn't require the original image for evaluation. PIQE models rely on large datasets of human-rated images, which can be challenging to obtain. The model might not generalize well to different image types or degradation types. Human perception of image quality can vary, affecting the training data. These challenges related to data availability and model generalization need to be addressed for widespread adoption.

Different augmentation techniques might impact these metrics differently. Different metrics have varying degrees of correlation with human perception. Determining appropriate thresholds for PSNR or SSIM to identify excessive distortion can be challenging. Using multiple metrics can provide a more comprehensive assessment of image quality. By monitoring these metrics during augmentation, practitioners can prevent excessive image degradation and maintain data quality.

5.1.4 Subjective Assessment

Human experts play a critical role in evaluating the quality and effectiveness of image augmentation techniques [25]. While objective metrics provide quantitative assessments, human perception offers invaluable qualitative insights. Human experts can assess how realistically augmented images mimic real-world scenarios. They can evaluate if augmentations improve model performance in terms of generalization and robustness. Humans can identify visual artifacts or distortions introduced by augmentation. Feedback on the overall visual appeal of augmented images is essential for applications like augmented reality or image editing.

Human evaluations provide valuable insights into the perceptual quality and effectiveness of augmented images. Here are some specific methods. The paired comparison method presents pairs of images, one original and one augmented, to human raters [26]. Raters are asked to choose the image that appears more natural or realistic. This method is simple to implement and can reveal subtle differences between images but can be time-consuming for large datasets, and results might be influenced by order effects. The ranking method presents a set of images, including originals and augmentations, to human raters and asks them to rank the images based on quality or realism. This method allows for the comparison of multiple images simultaneously. This method can be challenging for raters to differentiate between subtle differences in a large set of images. The rating scales method provides human raters with a rating scale (e.g., Likert scale) to assess image quality on different dimensions (naturalness, distortion, clarity). This method enables quantitative analysis of image quality but can be influenced by rater bias and the choice of rating scale. The crowdsourcing method utilizes online platforms to recruit a large number of participants for image evaluations. This method is cost-effective and can gather a diverse range of opinions but the quality control can be challenging, and participant motivation might vary. The expert evaluation method involves domain experts (e.g., radiologists, and photographers) to assess the impact of augmentations on image interpretation. This method provides in-depth insights into specific application areas but can be time-consuming and expensive. The preference tests involve conducting user studies to compare different augmentation techniques based on human perception. This approach helps identify the augmentation methods that are most visually appealing, effective, and aligned with user expectations. For this, first, a representative set of images from the dataset is selected. Then different augmentation techniques are applied to the selected images, creating multiple augmented versions of each image. Next, a group of participants with diverse backgrounds is recruited to ensure representativeness. Then, the participants are presented with pairs or groups of augmented images and ask them to express their preferences. Lastly, participant responses are analyzed to identify the most preferred augmentation techniques. For example, to compare the effectiveness of rotation, flipping, and color jittering, participants are presented with three versions of the same image: original, rotated, flipped, and color-jittered. Participants are asked to rank the images based on their preference for naturalness and image quality. A combination of these methods can provide a more comprehensive evaluation of augmented images. For example, paired comparisons can be used to identify subtle differences, while rating scales can provide quantitative data. By carefully designing human evaluation studies and considering the specific goals of the image augmentation project, valuable insights can be obtained.

Human evaluation studies, while valuable, present several challenges [27]. Different raters may have varying perceptions of image quality, leading to inconsistent results. The same rater might provide different ratings over time due to fatigue or changing criteria. Conducting large-scale human evaluations can be time-consuming and expensive. Recruiting and managing a sufficient number of human raters can be challenging. Personal preferences or expectations can influence rater judgments. Raters might focus on specific image attributes, ignoring others. Establishing clear criteria for image quality can be subjective and challenging. Selecting appropriate reference images for comparison can be difficult. Ensuring the well-being of human participants is crucial, especially when dealing with sensitive content. Addressing these challenges requires careful study design, rigorous data analysis, and the use of appropriate statistical methods. Despite these challenges, human evaluation remains an essential component of image augmentation research, providing valuable insights that complement objective metrics.

5.2   Combining Deep Learning and Traditional Techniques

By synergizing deep learning and traditional techniques, we can achieve a more comprehensive and robust evaluation of image augmentation [28]. Combine deep learning-based metrics (e.g., FID, KID) with traditional metrics (PSNR, SSIM) for a more comprehensive evaluation. By incorporating the perceptual loss functions into deep learning models, we can assess the perceptual quality of augmented images. We can use generative models to evaluate the diversity and realism of augmented images. For example, we can train a GAN to generate images similar to the original dataset. Then calculate FID and KID between the original, augmented, and GAN-generated images. After that, we can use a perceptual loss function to compare the feature representations of original and augmented images. Next, human evaluations can be conducted to assess the visual quality and realism of augmented images.

Combining deep learning and traditional techniques for image augmentation evaluation presents several challenges [29]. Both deep learning models and traditional metrics often require large datasets for accurate evaluation. The quality of the training data for deep learning models significantly impacts the performance of the combined approach. Training and evaluating deep learning models can be computationally expensive. Extracting features for traditional metrics might also be computationally intensive. Some deep learning metrics might correlate highly with traditional metrics, providing limited additional information. Identifying the complementary strengths of different metrics is crucial for effective combination. Deep learning models can be difficult to interpret, making it challenging to understand the reasons behind evaluation results. Combining deep learning with human evaluation requires careful consideration of human perception and biases. Training deep learning models and running computationally intensive traditional metrics can be resource-intensive, requiring high-performance hardware. Addressing these challenges requires careful experimentation and the selection of appropriate techniques for specific applications.

Some of the techniques to address these challenges are as follows. We can use GPUs or TPUs for faster training and inference [30]. One key strategy to overcome the computational challenges associated with deep learning, particularly when dealing with complex models or large datasets, is to use specialized hardware. GPUs (Graphics Processing Units) and TPUs (Tensor Processing Units) are specifically designed to accelerate the matrix operations that are fundamental to deep learning algorithms. These powerful processors offer significant speedups compared to traditional CPUs, enabling faster training and inference times. GPUs, initially designed for graphics rendering, have been adapted for parallel processing tasks, making them highly effective for training deep neural networks. TPUs, developed by Google, are specifically designed for machine learning workloads, offering even greater performance gains for deep learning tasks. By utilizing these specialized hardware accelerators, researchers and practitioners can significantly reduce training times and improve the overall efficiency of their deep learning models. This allows for faster experimentation, more rapid model development, and ultimately, more efficient deployment of deep learning solutions.

The cloud-based platforms can be utilized for large-scale computations and data storage [31]. Cloud-based platforms provide a powerful solution for overcoming the computational challenges associated with deep learning, particularly when dealing with large datasets and complex models. These platforms offer on-demand access to high-performance computing resources, including powerful GPUs and TPUs, enabling researchers and developers to scale their deep learning workloads effectively. By using cloud computing, researchers can access the necessary computational power without the need for significant upfront investments in hardware. This flexibility allows for on-demand scaling of resources based on the specific needs of the project, making it cost-effective and efficient to train and deploy deep learning models. Furthermore, cloud platforms provide access to a wide range of tools and services that simplify the process of developing, deploying, and managing deep learning models, such as pre-built machine learning frameworks, data storage and management services, and model deployment tools. By utilizing the scalability and flexibility of cloud computing, researchers can overcome the computational challenges associated with deep learning and accelerate the development and deployment of innovative AI solutions.

The workload can be distributed across multiple machines for parallel processing [32]. To address the computational demands of deep learning, particularly when dealing with large datasets or complex models, distributing the workload across multiple machines for parallel processing offers significant advantages. By dividing the computational task into smaller sub-tasks and executing them concurrently on multiple machines, we can significantly accelerate the training process.

The pre-trained models can be used to reduce the amount of training data required [33]. Using pre-trained models can significantly reduce the amount of training data required for a specific task. These pre-trained models, often trained on massive datasets like ImageNet, have already learned general features and representations from a wide range of images. By fine-tuning these pre-trained models on a smaller, task-specific dataset, we can use the knowledge acquired during the pre-training phase. This transfer learning approach allows us to achieve high performance with significantly less training data than would be required to train a model from scratch. For example, a pre-trained model for image classification, such as ResNet or Inception, can be fine-tuned for a specific task like classifying medical images. By initializing the model with the pre-trained weights and then training it on the smaller medical image dataset, we can use the general image understanding capabilities learned during pre-training. This approach not only reduces the amount of training data required but also accelerates the training process and improves the overall performance of the model.

By adopting these strategies, researchers can effectively address the challenges associated with combining deep learning and traditional techniques for image augmentation evaluation.

5.3   Optimization Techniques for Deep Learning-Based Image Augmentation

Optimization of image augmentation techniques involves selecting the most effective augmentation strategies and their parameters to maximize model performance.

Hyperparameter tuning [34] is a critical step in optimizing image augmentation pipelines. It involves systematically exploring different combinations of augmentation parameters to find the best configuration for a given task. Grid search is a hyperparameter tuning method that exhaustively evaluates all possible combinations of hyperparameters within a specified range. It creates a grid of parameter values and trains a model for each combination. While this approach guarantees finding the optimal hyperparameters within the specified grid, it can be computationally expensive, especially for high-dimensional hyperparameter spaces. This is particularly problematic for image augmentation, where multiple hyperparameters often need to be tuned simultaneously. Despite its computational cost, grid search provides a baseline for comparison with other hyperparameter optimization techniques. Random search is an alternative to grid search that explores the hyperparameter space by randomly sampling values from specified distributions. This approach often proves more efficient than grid search, especially when dealing with high-dimensional hyperparameter spaces. By randomly selecting combinations of hyperparameters, random search can discover promising regions within the search space more quickly. However, it might not guarantee to find the global optimum and can be less deterministic than a grid search. Despite these limitations, random search is widely used in practice due to its computational efficiency and ability to explore a larger portion of the hyperparameter space. Often more efficient than grid search, especially for high-dimensional hyperparameter spaces. Bayesian optimization is a probabilistic approach to hyperparameter tuning that builds a probabilistic model of the objective function (model performance) based on previous evaluations. This model is used to intelligently explore the hyperparameter space, focusing on regions with a high probability of improvement. By iteratively refining the probabilistic model, Bayesian optimization efficiently identifies promising hyperparameter configurations. This method is particularly effective when the evaluation of each hyperparameter combination is computationally expensive, as in the case of training deep learning models with extensive image augmentation pipelines. Compared to grid and random search, Bayesian optimization often converges to optimal hyperparameters faster, requiring fewer model evaluations. Gradient-based optimization directly optimizes hyperparameters by calculating gradients of the objective function (e.g., validation loss) concerning the hyperparameters. This approach treats hyperparameters as learnable parameters, similar to model weights. By iteratively updating hyperparameters based on the gradient information, it's possible to find optimal configurations efficiently. However, gradient-based optimization for hyperparameters is computationally expensive and requires specialized techniques to handle discrete hyperparameters. Additionally, it might not be suitable for all types of hyperparameters or optimization problems. Despite these challenges, gradient-based methods offer the potential for significant improvements in hyperparameter tuning efficiency when applicable. Early stopping is a regularization technique that helps prevent overfitting by terminating the training process when the model's performance on a validation set stops improving. This can be applied to hyperparameter tuning by stopping the search process for a particular hyperparameter combination if the model's performance plateaus. Some hyperparameters interact with each other, making the optimization process more complex. Hyperparameter interaction refers to the phenomenon where the effect of one hyperparameter is influenced by the values of other hyperparameters. In the context of image augmentation, this complexity arises due to the interconnectedness of augmentation techniques. For instance, the optimal rotation angle might vary depending on the scale factor or shear applied to an image. Similarly, the impact of color jitter might be influenced by the level of noise added. These interactions can make the hyperparameter optimization process challenging, as it requires exploring a vast and complex hyperparameter space to identify the optimal configuration. To address this, techniques like Bayesian optimization, which can model these interactions, can be employed. Hyperparameter tuning for image augmentation can be computationally demanding due to the iterative process of training and evaluating models with different hyperparameter configurations. The exploration of a vast hyperparameter space, coupled with the need to train multiple models, necessitates substantial computational resources. Factors such as the size of the dataset, the complexity of the augmentation techniques, and the number of hyperparameters to tune contribute to the overall computational cost. Access to high-performance computing infrastructure, including GPUs or TPUs, is often essential for efficient hyperparameter tuning. The appropriate evaluation metrics can be selected to assess the impact of hyperparameter tuning on model performance.

Augmentation policy learning [35] is a paradigm shift in data augmentation. Instead of manually designing augmentation pipelines, this approach learns optimal augmentation strategies directly from data. Reinforcement learning (RL) can be used to create adaptive image augmentation strategies. Here, an RL agent acts as the "augmentation policy," automatically selecting the most effective augmentation techniques for each training image. The training dataset and model itself become the "environment" for the agent. The agent receives a "reward" based on the model's performance on a validation set. High performance translates to high rewards. Through an iterative process, the agent learns to choose augmentations that consistently improve the model's performance, leading to an adaptive and data-driven approach to image augmentation. Meta-learning for augmentation involves a hierarchical learning process. At the outer level, a meta-learner acquires knowledge about effective augmentation strategies across various datasets and model architectures. This knowledge is then applied to the inner loop, where a base model is trained on a specific task using the learned augmentation policy. The meta-learner model learns a general-purpose augmentation strategy that can be adapted to different scenarios. The base learner is a standard model (e.g., CNN, RNN) trained on a specific dataset using the augmentation policy provided by the meta-learner. The base model is trained on a specific task using the current augmentation policy. The meta-learner updates its augmentation policy based on the performance of the base model on the validation set. By iteratively refining the augmentation policy, meta-learning aims to discover optimal augmentation strategies that can be applied to new tasks with minimal fine-tuning.

Augmentation policy learning offers several key advantages. It enables adaptive augmentation strategies, and tailoring techniques to specific image content and model characteristics. By learning optimal augmentation combinations directly from data, it often surpasses the efficiency of traditional grid or random search methods. This approach can lead to more generalized augmentation policies, making them applicable across various datasets and model architectures. Additionally, augmentation policy learning can help extract maximum value from limited datasets by optimizing data augmentation strategies.

Augmentation policy learning, while promising, faces several challenges. Firstly, the design of effective reward functions that accurately reflect model performance is non-trivial. Secondly, the exploration–exploitation trade-off, balancing the need to discover new augmentation strategies with exploiting known good policies, is crucial. Thirdly, computational costs associated with training reinforcement learning or meta-learning agents can be substantial. Additionally, ensuring that learned policies generalize well to unseen data remains a challenge.

Addressing the challenges of augmentation policy learning requires a combination of methodological advancements and computational resources. Carefully crafting reward functions that accurately reflect model performance is crucial. Incorporating multiple performance metrics can provide a more comprehensive evaluation. Employing techniques like epsilon-greedy, upper confidence bound (UCB), or Thompson sampling can help balance the exploration of new augmentations with the exploitation of known good policies [36]. Utilizing efficient reinforcement learning algorithms, such as Proximal Policy Optimization (PPO) or Deep Q-Networks (DQN), can accelerate training and improve sample efficiency [37]. Using pre-trained models or knowledge transfer from other tasks can help address data scarcity issues. Investing in high-performance computing infrastructure is essential for training complex augmentation policies. By combining these strategies, researchers can develop robust and effective augmentation policy learning systems.

Combining multiple augmentation techniques can significantly enhance the diversity and robustness of a training dataset [38]. However, the order in which these augmentations are applied can impact the final results. Combining multiple augmentation techniques can often yield better results than applying them individually. For instance, combining rotation with cropping exposes the model to a wider range of object orientations and scales, enhancing its ability to generalize. Similarly, applying color jittering after a geometric transformation can introduce additional diversity without compromising object integrity. The order in which augmentations are applied can also influence their effectiveness, as certain combinations might produce unexpected or beneficial outcomes. Understanding these synergistic effects is crucial for optimizing image augmentation pipelines and improving model performance. The order in which augmentation techniques are applied can significantly impact the resulting image and, consequently, the model's performance. For instance, applying a rotation to an image before adjusting its color might lead to different color distributions compared to applying color augmentation first. This is because color transformations might be affected by the spatial rearrangement of pixels caused by rotation. Similarly, applying cropping before noise addition can result in different noise patterns compared to the reverse order. Understanding these dependencies is crucial for optimizing augmentation pipelines and achieving desired image transformations. Experimentation is a cornerstone of effective image augmentation. Exploring various combinations and sequences of augmentation techniques is crucial for identifying the optimal configuration for a specific task. By systematically testing different approaches, practitioners can uncover synergistic effects between augmentations, such as how combining rotation with cropping can expose the model to diverse object orientations and scales. Understanding the impact of augmentation order is equally important, as applying transformations in different sequences can produce varying results. This iterative process of experimentation and refinement is essential for maximizing the benefits of image augmentation and achieving optimal model performance.

The **data-centric augmentation** is a targeted approach to augmentation that involves focusing on the most challenging data samples for the model [39]. By identifying and augmenting misclassified or low-confidence samples, practitioners can effectively address the model's weaknesses. This strategy helps to improve performance on specific image regions or object categories. For instance, in object detection, augmenting images with small or occluded objects can enhance the model's ability to detect these challenging instances. By concentrating augmentation efforts on areas where the model struggles, there is a greater potential for performance improvement compared to applying augmentations uniformly to the entire dataset.

Augmentation scheduling involves strategically varying the intensity or frequency of augmentations during the training process [40]. By dynamically adjusting augmentation parameters, it helps prevent overfitting and improve generalization. For instance, applying stronger augmentations in the early stages of training can expose the model to diverse data, while gradually reducing augmentation intensity later can refine the model's learning. This approach mimics the learning process in humans, where early exposure to diverse stimuli is followed by a focus on fine-tuning details.

By carefully optimizing augmentation techniques and evaluating their impact, researchers can significantly improve the performance of deep learning models.

5.4 Summary

In this chapter different strategies for evaluating and optimizing deep learning image augmentation are discussed. Effectively assessing the impact of image augmentation on model performance requires a multifaceted approach. Traditional metrics such as accuracy, precision, recall, F1-score, and confusion matrices are foundational for classification tasks. For object detection, metrics like mAP, IoU, and its variants

(GIoU, DIoU, CIoU) are employed to evaluate localization and classification accuracy. In image segmentation, pixel accuracy, mIoU, Dice coefficient, and Panoptic Quality (PQ) are crucial. To measure image quality and diversity, metrics like PSNR, SSIM, FID, KID, Shannon entropy, and differential entropy are employed. Human perception, assessed through paired comparisons, ranking, and rating scales, provides valuable qualitative insights. Integrating deep learning techniques, such as using GANs or autoencoders, can enhance evaluation by offering additional perspectives on image quality and diversity. Hyperparameter tuning techniques including grid search, random search, Bayesian optimization, and gradient-based methods optimize augmentation parameters. Augmentation policies learned through reinforcement learning or meta-learning offer adaptive strategies. Data-centric augmentation focuses on challenging samples, while augmentation scheduling varies augmentation intensity over time. By combining these approaches and considering computational resources, researchers can develop effective image augmentation pipelines that enhance model performance.

References

1. Gu, S., Pednekar, M., & Slater, R. (2019). Improve image classification using data augmentation and neural networks. *SMU Data Science Review, 2*(2), 1.
2. Elgendi, M., Nasir, M. U., Tang, Q., Smith, D., Grenier, J. P., Batte, C., Spieler, B., Leslie, W. D., Menon, C., Fletcher, R. R. & Nicolaou, S. (2021). The effectiveness of image augmentation in deep learning networks for detecting COVID-19: A geometric transformation perspective. *Frontiers in Medicine, 8*, 629134.
3. Hammoudi, K., Cabani, A., Slika, B., Benhabiles, H., Dornaika, F., & Melkemi, M. (2022). Superpixelgridmasks data augmentation: Application to precision health and other real-world data. *Journal of Healthcare Informatics Research, 6*(4), 442–460.
4. Pham, T. C., Doucet, A., Luong, C. M., Tran, C. T., & Hoang, V. D. (2020). Improving skin-disease classification based on customized loss function combined with balanced mini-batch logic and real-time image augmentation. *IEEE Access, 8*, 150725–150737.
5. Anwar, T., & Zakir, S. (2021, April). Effect of image augmentation on ECG image classification using deep learning. In *2021 international conference on artificial intelligence (ICAI)* (pp. 182–186). IEEE.
6. Prasanna, S., & El-Sharkawy, M. (2022, August). Improving mean average precision (mAP) of camera and radar fusion network for object detection using radar augmentation. In *Proceedings of seventh international congress on information and communication technology: ICICT 2022, London, Volume 4* (pp. 51–60). Springer Nature Singapore.
7. Budiarsa, R., Wardoyo, R., & Musdholifah, A. (2024). Face recognition with occluded face using improve intersection over union of region proposal network on Mask region convolutional neural network. *International Journal of Electrical & Computer Engineering (2088–8708), 14*(3).
8. Tong, C., Yang, X., Huang, Q., & Qian, F. (2022). NGIoU Loss: Generalized intersection over union loss based on a new bounding box regression. *Applied Sciences, 12*(24), 12785.
9. Yu, Z., Yang, D., Wu, W., Wang, Y., & Luo, Y. (2022, December). Fast convergence detection algorithm of image small object based on distance intersection over union. In *2022 4th international conference on control and robotics (ICCR)* (pp. 330–336). IEEE.

10. Hao, W., Zhang, L., Han, M., Zhang, K., Li, F., Yang, G., & Liu, Z. (2023). YOLOv5-SA-FC: A novel pig detection and counting method based on shuffle attention and focal complete intersection over union. *Animals, 13*(20), 3201.

11. Shorten, C., & Khoshgoftaar, T. M. (2019). A survey on image data augmentation for deep learning. *Journal of big data, 6*(1), 1–48.

12. Xu, M., Yoon, S., Fuentes, A., & Park, D. S. (2023). A comprehensive survey of image augmentation techniques for deep learning. *Pattern Recognition, 137*, 109347.

13. Rezatofighi, H., Tsoi, N., Gwak, J., Sadeghian, A., Reid, I., & Savarese, S. (2019). Generalized intersection over union: A metric and a loss for bounding box regression. In *Proceedings of the IEEE/CVF conference on computer vision and pattern recognition* (pp. 658–666).

14. Asad, M. H., & Bais, A. (2020). Weed detection in canola fields using maximum likelihood classification and deep convolutional neural network. *Information Processing in Agriculture, 7*(4), 535–545.

15. Saha, S., Hoyer, L., Obukhov, A., Dai, D., & Van Gool, L. (2023). EDAPS: Enhanced domain-adaptive panoptic segmentation. In *Proceedings of the IEEE/CVF international conference on computer vision* (pp. 19234–19245).

16. Noguchi, S., Nishio, M., Yakami, M., Nakagomi, K., & Togashi, K. (2020). Bone segmentation on whole-body CT using convolutional neural network with novel data augmentation techniques. *Computers in Biology and Medicine, 121*, 103767.

17. Yu, Y., Zhang, W., & Deng, Y. (2021). Frechet inception distance (fid) for evaluating gans. *China University of Mining Technology Beijing Graduate School, 3*.

18. Muhammad, A., Salman, Z., Lee, K., & Han, D. (2023). Harnessing the power of diffusion models for plant disease image augmentation. *Frontiers in Plant Science, 14*, 1280496.

19. Nouara, B., & Bilal, M. (2024, October). Random optimization and entropy-based data augmentation for image classification and analysis "ROEDA". In *2024 International conference on advances in electrical and communication technologies (ICAECOT)* (pp. 1–6). IEEE.

20. Kim, E., Kim, J., Lee, H., & Kim, S. (2021). Adaptive data augmentation to achieve noise robustness and overcome data deficiency for deep learning. *Applied Sciences, 11*(12), 5586.

21. Dwivedi, P., Padhi, S., Chakraborty, S., & Raikwar, S. C. (2024). Severity wise COVID-19 X-ray image augmentation and classification using structure similarity. *Multimedia Tools and Applications, 83*(10), 30719–30740.

22. Zhang, L., Zhang, L., Mou, X., & Zhang, D. (2011). FSIM: A feature similarity index for image quality assessment. *IEEE Transactions on Image Processing, 20*(8), 2378–2386.

23. Gong, C., Wang, D., Li, M., Chandra, V., & Liu, Q. (2021). Keepaugment: A simple information-preserving data augmentation approach. In *Proceedings of the IEEE/CVF conference on computer vision and pattern recognition* (pp. 1055–1064).

24. Eybposh, M. H., Cai, C., Moossavi, A., Rodriguez-Romaguera, J., & Pégard, N. C. (2024). ConIQA: A deep learning method for perceptual image quality assessment with limited data. *Scientific Reports, 14*(1), 20066.

25. Garcea, F., Serra, A., Lamberti, F., & Morra, L. (2023). Data augmentation for medical imaging: A systematic literature review. *Computers in Biology and Medicine, 152*, 106391.

26. Hodosh, M., Young, P., & Hockenmaier, J. (2013). Framing image description as a ranking task: Data, models and evaluation metrics. *Journal of Artificial Intelligence Research, 47*, 853–899.

27. Stephanidis, C., Salvendy, G., Antona, M., Chen, J. Y., Dong, J., Duffy, V. G., Fang, X., Fidopiastis, C., Fragomeni, G., Fu, L. P. & Zhou, J. (2019). Seven HCI grand challenges. *International Journal of Human–Computer Interaction, 35*(14), 1229–1269.

28. Mumuni, A., & Mumuni, F. (2022). Data augmentation: A comprehensive survey of modern approaches. *Array, 16*, 100258.

29. Nagaraju, M., Chawla, P., & Kumar, N. (2022). Performance improvement of Deep Learning Models using image augmentation techniques. *Multimedia Tools and Applications, 81*(7), 9177–9200.

30. Kljucaric, L., & George, A. D. (2023). Deep learning inferencing with high-performance hardware accelerators. *ACM Transactions on Intelligent Systems and Technology, 14*(4), 1–25.

31. Cai, H., Xu, B., Jiang, L., & Vasilakos, A. V. (2016). IoT-based big data storage systems in cloud computing: Perspectives and challenges. *IEEE Internet of Things Journal, 4*(1), 75–87.
32. Rashid, Z. N., Zebari, S. R., Sharif, K. H., & Jacksi, K. (2018, October). Distributed cloud computing and distributed parallel computing: A review. In *2018 International Conference on Advanced Science and Engineering (ICOASE)* (pp. 167–172). IEEE.
33. Han, X., Zhang, Z., Ding, N., Gu, Y., Liu, X., Huo, Y., Qiu, J., Yao, Y., Zhang, A., Zhang, L., & Zhu, J. (2021). Pre-trained models: Past, present and future. *AI Open, 2*, 225–250.
34. Cejudo Grano de Oro, J. E., Koch, P. J., Krois, J., Garcia Cantu Ros, A., Patel, J., Meyer-Lueckel, H., & Schwendicke, F. (2022). Hyperparameter tuning and automatic image augmentation for deep learning-based angle classification on intraoral photographs—A retrospective study. *Diagnostics, 12*(7), 1526.
35. Zhou, F., Li, J., Xie, C., Chen, F., Hong, L., Sun, R., & Li, Z. (2021, May). Metaaugment: Sample-aware data augmentation policy learning. In *Proceedings of the AAAI conference on artificial intelligence* (Vol. 35, No. 12, pp. 11097–11105).
36. Tao, J. (2023). *Improving decision making through learning*. The University of Wisconsin-Madison.
37. Bondre, S. V., Thakre, B., Yadav, U., & Bondre, V. D. (2024). Deep reinforcement learning algorithms: A comprehensive overview. *Deep Reinforcement Learning and Its Industrial Use Cases: AI for Real-World Applications*, 51–73.
38. Rebuffi, S. A., Gowal, S., Calian, D. A., Stimberg, F., Wiles, O., & Mann, T. A. (2021). Data augmentation can improve robustness. *Advances in Neural Information Processing Systems, 34*, 29935–29948.
39. Bhatt, N., Bhatt, N., Prajapati, P., Sorathiya, V., Alshathri, S., & El-Shafai, W. (2024). A data-centric approach to improve performance of deep learning models. *Scientific Reports, 14*(1), 22329.
40. Liang, W., Liang, Y., & Jia, J. (2023). MiAMix: Enhancing image classification through a multi-stage augmented mixed sample data augmentation method. *Processes, 11*(12), 3284.

Chapter 6
The Future of Deep Learning Image Augmentation

Image augmentation has proven to be a cornerstone of successful deep learning models in computer vision. By artificially expanding the training dataset, augmentation techniques enhance model robustness, generalization, and performance. However, the field is continually evolving, with exciting advancements on the horizon. The future of deep learning image augmentation lies in pushing the limits of current methods and exploring novel approaches [1]. This chapter will delve into emerging trends and promising directions, including the integration of generative models, the development of more sophisticated augmentation policies, and the exploration of domain-specific augmentation strategies.

6.1 AutoAugment: A Reinforcement Learning Approach

AutoAugment is a pioneering work in automated image augmentation [2]. It uses Reinforcement Learning (RL) to discover effective augmentation policies directly from the data. AutoAugment frames the problem of finding the best augmentation policy as a discrete search problem. Instead of manually defining augmentation techniques, it treats the selection and application of augmentations as a decision-making process. Essentially, AutoAugment aims to learn an optimal "augmentation policy" from the image itself [3]. This policy defines:

(a) Which augmentation operations to apply: The "Which augmentation operations to apply" aspect of AutoAugment defines the search space for the augmentation policy. This involves carefully selecting a set of candidate augmentation operations that are relevant to the image domain and the specific task. Common operations include geometric transformations like rotation (Rotating the image by a specified angle), translation (Shifting the image horizontally or vertically), shearing (Skewing the image along one or both axes), scaling (Resizing the

image), and flipping (Flipping the image horizontally or vertically); color transformations such as brightness (Increasing or decreasing the overall brightness
of the image), contrast (Increasing or decreasing the contrast of the image),
saturation (Increasing or decreasing the color saturation), and hue adjustments
(Shifting the hue of the image); noise augmentation like Gaussian noise (Adding
Gaussian noise to the image) and salt and pepper noise (Adding salt and pepper
noise to the image); and other techniques like cutout (Removing rectangular
patches from the image) and random erasing (Randomly erasing regions within
the image). This set of candidate operations provides the building blocks for
the controller RNN to explore and discover effective augmentation policies by
selecting and combining these operations in various ways.

(b) Probability of applying each operation: The "probability of applying each operation" aspect introduces a crucial element of stochasticity to AutoAugment.
Instead of deterministically applying a fixed set of augmentations to every
image, the model learns to probabilistically select which operations to apply.
For example, An augmentation policy might dictate that "rotation" is applied
with a probability of 0.7. This means that in 70% of cases, the image will be
randomly rotated, while in the remaining 30%, no rotation will occur. Similarly,
"shearing" might be applied with a probability of 0.3, and "color jittering" with
a probability of 0.8. This probabilistic approach offers several advantages. It
allows for a wider range of possible augmentations to be explored, increasing
the diversity of the training data. The probabilities can be adapted based on the
specific features of the data and the model's performance. By not applying all
augmentations to every image, the model is less likely to overfit specific transformations. This probabilistic selection of operations meaningfully improves the
flexibility and effectiveness of AutoAugment, enabling the discovery of more
robust and data-efficient augmentation strategies.

(c) The magnitude of each operation: The "magnitude of each operation" refers
to the intensity or degree to which an augmentation is applied. This parameter significantly influences the level of transformation and the resulting image
characteristics. For Example, the magnitude of rotation would be the angle of
rotation, such as 15 degrees, 30 degrees, or 45 degrees. The magnitude of scaling
could be the scaling factor, such as 0.8 (reducing the image size by 20%) or 1.2
(increasing the image size by 20%). The magnitude of shearing could be the
shear factor, determining the degree of skewing. The magnitude of color jitter
could be the range of brightness, contrast, saturation, or hue adjustments. For
example, brightness might be adjusted by a factor between -0.2 and 0.2. The
magnitude of noise addition could be the level of noise added, such as the standard deviation of Gaussian noise or the density of salt and pepper noise. By
varying the magnitude of each operation, AutoAugment explores a wider range
of transformations and learns to select the most effective levels for different
images and tasks. For example, a small rotation might be beneficial for some
images, while a larger rotation might be necessary for others. By learning the
optimal magnitudes for each operation, AutoAugment can significantly improve
the diversity and quality of augmented data.

The required Python code of the above can be the following:

```python
import random
def apply_augmentations(image):
    """

    Applies a set of augmentations to the input image with specified
probabilities and magnitudes.
    Args:
        image: The input image.
    Returns:
        The augmented image.
    """
    operation_probabilities = {
        'rotate': 0.7,   # 70% chance of applying rotation
        'shear': 0.3,    # 30% chance of applying shear
        'color_jitter': 0.8,  # 80% chance of applying color jitter
        'horizontal_flip': 0.5   # 50% chance of applying horizontal flip
    }
    for operation, probability in operation_probabilities.items():
        if random.random() < probability:
            if operation == 'rotate':
                # Randomly select a rotation angle (magnitude) between -
15 and 15 degrees
                angle = random.uniform(-15, 15)
                image = rotate(image, angle)
            elif operation == 'shear':
                # Randomly select a shear factor between -0.2 and 0.2
                shear = random.uniform(-0.2, 0.2)
                image = shear(image, shear)
            elif operation == 'color_jitter':
                # Apply random brightness, contrast, saturation, and hue
adjustments
                image = color_jitter(image)
            elif operation == 'horizontal_flip':
                image = flip(image, 'horizontal')
    return image
```

This code snippet demonstrates the key concepts. The `operation_`
`probabilities` dictionary specifies the probability of applying each augmentation operation. The code uses `random.random()` to determine whether to apply each operation based on its probability. For each operation, a random value within a specified range is selected to determine the magnitude of the transformation (e.g., rotation angle, shear factor). This example provides a basic framework for implementing probabilistic augmentation policies in AutoAugment. In practice, more complex policies and a wider range of augmentation operations can be explored.

6.1.1   Controller

The controller in AutoAugment is a crucial component, acting as the "brain" that generates augmentation policies [4]. Specifically, a Recurrent Neural Network (RNN), often an LSTM (Long Short-Term Memory) network, serves as this controller. The RNN operates sequentially, making a series of decisions to define the augmentation policy. At each step, the RNN predicts the following.

(a) The next augmentation operation: The RNN chooses from a predefined set of candidate operations (e.g., rotation, shear, color jitter, etc.). This involves making a categorical prediction, often implemented as a softmax layer. Imagine a scenario with five possible operations: Rotation, Shear, Color Jitter, Horizontal Flip, and No Operation (i.e., no augmentation). The RNN processes information about the current image (or potentially previous augmentation decisions) and generates a vector of scores, one for each operation. To convert these scores into probabilities, a softmax function is applied. Softmax ensures that each operation is assigned a probability between 0 and 1 and the sum of probabilities for all operations equals 1. This results in a probability distribution over the set of candidate operations. For example, the RNN might output the following probabilities: Rotation (0.3), Shear (0.1), Color Jitter (0.5), Horizontal Flip (0.1), and No Operation (i.e., no augmentation) (0.0). Based on this distribution, the RNN selects the next operation to apply. In this case, "Color Jitter" has the highest probability (0.5), so it would be the most likely operation to be chosen. By using the SoftMax function, the RNN ensures that the probabilities of all operations sum to 1, providing a well-defined probability distribution for selecting the next augmentation operation. This mechanism allows the RNN to learn to effectively prioritize different operations based on the specific features of the image and the anticipated augmentation policy.

(b) The probability of applying the selected operation: The RNN predicts the probability of applying the chosen operation. This allows for stochasticity in the augmentation process, where some operations are applied more frequently than others.

(c) The magnitude of the operation: For each selected operation, the RNN predicts
the magnitude (e.g., the angle of rotation, and the intensity of color jitter). This
is often achieved through regression or by predicting values within a specified
range.

The RNN processes these decisions sequentially, effectively generating a chain
of augmentation operations with associated probabilities and magnitudes. This
sequential nature allows the RNN to learn complex dependencies between different
operations and their effects on the final image.

6.1.2 Reinforcement Learning Loop

The reinforcement learning loop in AutoAugment involves a cyclical process of
policy generation, data augmentation, model training, and policy improvement [5,
6].

(a) Policy generation: The controller RNN in AutoAugment generates an augmentation policy through a sequential decision-making process. Firstly, the RNN
selects the next augmentation operation to apply. It does this by considering a
predefined set of candidate operations, such as rotation, shearing, color jittering,
and horizontal flipping. To make this selection, the RNN employs a softmax
layer, which outputs a probability distribution over the set of candidate operations. This distribution reflects the likelihood of each operation being chosen at
that particular step in the policy generation process. By using a softmax layer,
the RNN ensures that the probabilities for all operations sum to one, providing
a well-defined probability distribution for the next augmentation step.
(b) Data Augmentation: The generated augmentation policy is then applied to transform the training data. Each image undergoes a series of augmentations determined by the policy. The policy dictates which operations are applied, with
probabilities controlling the likelihood of each operation. The magnitude of
each applied operation, such as the rotation angle or color jitter intensity, is
also determined by the policy. This process results in a diverse set of augmented
images for each original image, revealing the model to a wide range of variations
and enhancing its ability to learn robust and generalizable representations.
(c) Child Model Training: The child model, typically a convolutional neural
network (CNN) for image tasks, is trained on the augmented dataset generated by the current augmentation policy. This training process involves standard
deep learning techniques. The model is iteratively updated using algorithms like
stochastic gradient descent (SGD) or its variants (e.g., Adam, RMSprop) to minimize the loss function, such as cross-entropy loss for classification. Backpropagation is employed to efficiently calculate the gradients and apprise the model's
weights. To improve training stability and prevent overfitting, techniques like

mini-batch training, dropout, weight decay, and even data augmentation at the child model level (e.g., random cropping, flipping) are often incorporated.

(d) Performance Evaluation: The trained child model is then rigorously evaluated on a held-out validation set, a portion of the data that was not used during training. This evaluation offers a crucial assessment of the child model's performance. Metrics such as accuracy, precision, recall, F1-score, and area under the ROC curve (AUC) are commonly used to measure the model's ability to correctly classify instances. By assessing the child model's performance on this unseen data, we can assess the effectiveness of the augmentation policy in improving the model's generalization ability. This performance score serves as a crucial signal for the next step: updating the controller RNN

(e) Reward Calculation: The child model's performance on the validation set serves as the crucial reward signal for the controller RNN. If the child model attains high accuracy on the validation set, it indicates that the augmentation policy generated by the controller was effective in improving the model's generalization ability. This high performance translates into a high reward signal for the controller RNN. Conversely, if the child model performs poorly on the validation set, it suggests that the current augmentation policy is not effective. This low performance results in a lower reward signal for the controller RNN. By associating the child model's performance with a reward signal, the reinforcement learning framework incentivizes the controller RNN to generate augmentation policies that consistently improve the model's performance on the validation set. This feedback loop drives the learning process, allowing the controller to gradually refine its policy and discover increasingly effective augmentation strategies. The required function for this can be as follows.

```
def calculate_reward(child_model, validation_data):
    """

    Calculates the reward based on the child model's performance on the
    validation set.
    Args:
        child_model: The trained child model.
        validation_data: The validation dataset.
    Returns:
        The reward value (e.g., accuracy).
    """
    # Evaluate the child model on the validation data
    loss, accuracy = child_model.evaluate(validation_data)
    # Reward can be directly the accuracy
    reward = accuracy
    # Alternatively, we can use a scaled reward
    # reward = accuracy * 100
    return reward
```

The calculate_reward function receipts the trained child_model and the validation_data as input. It evaluates the child_model on the validation_data using the model.evaluate() method. This typically returns the loss and a performance metric (e.g., accuracy). The reward is initially assigned the accuracy value. The code provides an option to scale the reward by multiplying it with a constant (e.g., 100) to increase the magnitude of the reward signal. The calculated reward is returned to the reinforcement learning algorithm to update the controller RNN's policy. This is a simplified example. In practice, more sophisticated reward functions can be designed, such as incorporating multiple metrics (e.g., accuracy, precision, recall) or using a weighted combination of metrics. The selection of the reward function depends on the specific task and the desired evaluation criteria. This reward signal acts as the feedback mechanism in the reinforcement learning loop, guiding the controller RNN toward generating augmentation policies that consistently improve the child model's performance on the validation set.

(f) Policy Update: The controller RNN is updated using policy gradient methods, such as Proximal Policy Optimization (PPO), to maximize the expected reward. These methods aim to directly improve the policy function that maps states (e.g., image features) to actions (augmentation operations and their magnitudes). Policy gradient methods work by calculating the gradient of the expected reward concerning the controller RNN's parameters. This gradient indicates the direction in which the controller's parameters should be adjusted to increase the expected reward. PPO is a popular policy gradient method that introduces a "clipped surrogate objective" to constrain the policy updates. This constraint helps to stabilize training and prevent large policy updates that can lead to insta-bility. PPO calculates the ratio of probabilities of taking the same action under the new policy (π_{new}) and the old policy (π_{old}). This ratio often denoted as "$r(\theta)$", quantifies how much the policy has changed. PPO introduces a clipping factor, typically denoted as "ϵ" (epsilon). This factor defines a permissible range for the probability ratio: $[1 - \epsilon, 1 + \epsilon]$. If the probability ratio falls outside this range, it is clipped to the corresponding boundary. PPO then constructs a "clipped surrogate objective" function. This function takes the minimum of two values: (1) The original, unclipped objective function (which aims to maximize the expected reward), and (2) The clipped probability ratio multiplied by the advan-tage. The controller RNN's parameters are then updated using gradient descent to maximize the clipped surrogate objective. By clipping the probability ratio, PPO ensures that the policy updates are not too large, preventing drastic changes that could lead to instability and poor performance. This constraint allows for more stable and reliable policy learning, enabling the controller RNN to gradu-ally improve its augmentation strategies while maintaining a degree of control over the policy updates. The PPO objective function encourages the controller RNN to improve the policy while ensuring that the policy updates are not too drastic. This prevents the policy from deviating significantly from the previous policy, which can lead to instability and poor performance. PPO introduces a "clipped surrogate objective" to achieve this. This objective function aims to: (1) Maximize expected reward: Encourage the controller RNN to generate policies that lead to higher rewards (i.e., better child model performance). (2) Constrain policy updates: Prevent the policy from changing too much in a single update step. The core idea is to compare the probability of taking an action under the current policy (π_θ) to the probability of taking the same action under the previous policy (π_θ_old). If this probability ratio deviates significantly from 1, it indicates a large change in the policy. PPO introduces a "clipping" mechanism. If the probability ratio falls within a certain range (typically between $1 - \epsilon$ and $1 + \epsilon$, where ϵ is a small constant), the objective function remains unchanged. If the probability ratio exceeds this range, it is clipped to the upper or lower bound of the range. This clipping mechanism effectively limits the magnitude of policy updates, ensuring that the policy does not change too drastically in a single step. This constraint aids in stabilizing the training procedure and prevents the controller RNN from exploring overly aggressive and potentially unstable augmentation strategies. By using this clipped surrogate objective, PPO encour-ages the controller RNN to gradually improve the policy while maintaining a degree of stability and preventing large, potentially disruptive policy changes. This leads to more robust and reliable policy learning, ultimately resulting in more effective and stable augmentation strategies. The required code snippet of the policy update can be as follows:

```python
import tensorflow as tf
def ppo_loss(old_log_probs, log_probs, advantages, clip_epsilon=0.2):
  """
  Calculates the PPO loss.
  Args:
    old_log_probs: Log probabilities of actions under the old policy.
    log_probs: Log probabilities of actions under the new policy.
    advantages: Estimated advantages of taking each action.
    clip_epsilon: Clipping parameter for PPO.
  Returns:
    The PPO loss.
  """
  ratio = tf.exp(log_probs - old_log_probs)
  clipped_ratio = tf.clip_by_value(ratio,
                                   clip_value_min=1.0 - clip_epsilon,
                                   clip_value_max=1.0 + clip_epsilon)
  loss = -tf.reduce_mean(tf.minimum(ratio * advantages, clipped_ratio *
advantages))
  return loss
# Example usage:
# Assuming we have:
# - old_log_probs: Log probabilities of actions under the old policy.
#   Shape: (batch_size,)
# - log_probs: Log probabilities of actions under the new policy.
#   Shape: (batch_size,)
# - advantages: Estimated advantages of taking each action.
#   Shape: (batch_size,)
# Calculate the PPO loss
ppo_loss_value = ppo_loss(old_log_probs, log_probs, advantages)
# Use an optimizer (e.g., Adam) to update the controller RNN's parameters
# based on the calculated PPO loss.
optimizer = tf.keras.optimizers.Adam(learning_rate=0.001)
optimizer.minimize(lambda: ppo_loss_value,
                   var_list=controller_rnn.trainable_variables)
```

The `tf.clip_by_value` function clips the probability ratio to a specified range (in this example, between 0.8 and 1.2). This clipping operation prevents excessive policy changes. The PPO loss is calculated as the negative of the minimum between the original objective (ratio * advantages) and the clipped objective (clipped_ratio * advantages). This code snippet provides a basic implementation of the PPO loss calculation and demonstrates how to use it to update the controller RNN's parameters. This core concept forms the foundation for training the controller in AutoAugment and discovering effective augmentation policies. In a real-world implementation, we would need to handle batching, handle multiple steps in the RNN, and implement more sophisticated training techniques.

This iterative process continues, with the controller RNN continuously improving its ability to generate effective augmentation policies that cause better performance of the child model. This reinforcement learning loop enables AutoAugment to discover sophisticated and data-driven augmentation strategies that can significantly enhance the performance of deep learning models.

6.2   Interpretable Augmentation

Many common augmentation methods, like random rotations, cropping, and color jittering, can introduce significant variations in the input data, making it difficult to understand the model's decision-making process. This lack of interpretability can be a significant limitation, particularly in critical applications like medical image analysis or autonomous driving. Interpretable augmentation aims to address this challenge by developing augmentation techniques that are more transparent and easier to understand [7]. By designing augmentations that are more meaningful and less arbitrary, we can have a deeper understanding of the model's learning procedure and improve the trust and reliability of its predictions.

6.2.1   *Rule-Based Augmentation with Justifications*

Rule-based augmentation involves applying predefined transformations to the data based on specific rules or constraints [8]. Let's consider a medical imaging task involving chest X-rays for pneumonia detection. In this case, the Rule-Based Augmentation can have the following rule:

Rule 1: Simulate small random rotations: Rotate the X-ray image by a small random angle (e.g., between −5 and 5 degrees). This simulates slight variations in patient positioning during image acquisition.

Rule 2: Simulate slight intensity variations: Adjust the image intensity by a small random factor (e.g., between 0.9 and 1.1). This simulates variations in X-ray machine settings or differences in image acquisition techniques.

Rule 3: Simulate small random shifts: Shift the image slightly in the horizontal and vertical directions (e.g., by a few pixels). This simulates minor misalignments during image acquisition or processing.

This approach offers several advantages in terms of interpretability. Some of them are as follows.

Transparency: Rule-based augmentations are inherently transparent [9]. The rules governing the transformations are explicitly defined and easily understood. This makes it easier to understand how the augmented data differs from the original data and how these transformations might influence the model's learning process. For the above example, these rules are explicitly defined and easily understandable. We know exactly how the original image is being modified. This transparency allows researchers to (1) Understand the specific types of variations the model is being trained to handle, (2) Assess the impact of each augmentation rule on model performance and (3) Debug and troubleshoot potential issues related to specific augmentations. For example, if the model performs poorly after applying a particular rule (e.g., excessive intensity variations), the researcher can easily identify and adjust or remove that rule from the augmentation pipeline. This level of transparency is crucial for developing robust and reliable models, especially in critical applications like medical imaging.

Controllability: Rule-based approaches provide greater control over the augmentation process [10]. By carefully defining the rules and their parameters, researchers can confirm that the augmented data remains within a specific range of variations and avoids introducing unrealistic or undesirable artifacts. For example, in medical imaging, excessive rotations or intensity shifts might introduce unrealistic artifacts or distort anatomical structures. Rule-based augmentations can be designed to limit the extent of these transformations, ensuring that the augmented data remains within a plausible range of variations. In some cases, certain augmentations can introduce unrealistic or undesirable artifacts into the data. For example, excessive cropping might remove crucial information from the image. Rule-based approaches can be designed to avoid these situations by defining specific constraints and limitations on the augmentation process. This level of control is mainly important in critical applications where the accuracy and reliability of the model are paramount. By carefully defining and controlling the augmentation process, researchers can ensure that the augmented data remains relevant, realistic, and informative for the learning process.

Domain Knowledge Incorporation: Rule-based augmentation excels at incorporating domain-specific knowledge into the data augmentation process [11]. In medical imaging, this is particularly valuable. For example, in X-rays, slight variations in patient positioning can significantly impact image appearance. Rule-based augmentation can introduce small, realistic rotations and translations to simulate these variations. This helps the model learn to be robust to minor positioning errors that might occur during actual image acquisition. Different imaging devices and settings can produce images with varying levels of noise, contrast, and sharpness.

Rule-based augmentation can simulate these variations by adjusting image intensity, adding noise, or blurring the image slightly, mimicking the effects of different imaging equipment and settings. In some cases, augmentations can be designed to simulate physiological variations. For example, in cardiac imaging, augmentations could simulate variations in heart rate or breathing patterns, which can affect image appearance. By incorporating this domain-specific knowledge into the augmentation procedure, researchers can generate more realistic and challenging training data. This aids the model learn to be robust to real-world variations and improve its performance on unseen data.

Reproducibility: Rule-based augmentation techniques are highly reproducible due to the explicit and deterministic nature of the defined rules [12]. Unlike stochastic methods where the outcome might vary slightly with each application, rule-based methods are governed by precise rules and parameters. For example, a rule might state: "Rotate the image by a random angle between -10 and 10 degrees." This rule is clearly defined and can be consistently applied across different experiments and datasets. As long as the same rules and parameters are used, the same augmentations will be applied to the same input data. This ensures consistency and reproducibility across different experiments, allowing researchers to compare results more reliably and identify the true impact of different model architectures or training procedures. Researchers can easily share and reproduce the results of experiments using rule-based augmentations. By clearly documenting the rules and parameters used, other researchers can easily replicate the augmentation process and compare their results. This high level of reproducibility is crucial for scientific rigor and enables researchers to build upon the work of others, fostering progress and collaboration within the field of deep learning.

Explainability: Rule-based augmentations enhance the explainability of the model's predictions by providing insights into the model's sensitivity to specific types of variations [13]. By observing how the model performs with different rule-based augmentations, researchers can gain insights into the model's sensitivity to specific types of transformations. For example, if the model's performance significantly degrades when images are rotated by large angles, it suggests that the model might be overly sensitive to object orientation. Rule-based augmentations can help identify limitations and biases in the model. If the model struggles to generalize to images with specific types of noise introduced by an augmentation rule, it might indicate a weakness in the model's ability to handle noisy or degraded data. By understanding the model's sensitivity to different augmentations, researchers can refine the augmentation strategy to enhance the model's generalization ability. For instance, if the model struggles with small rotations, the augmentation pipeline can be adjusted to include more rotations within a specific range. Rule-based augmentations can help debug model behavior. If the model consistently misclassifies images after a specific type of augmentation, it might indicate a flaw in the model's architecture or training process.

The following code snippet demonstrates a simple example of rule-based augmentations. We can easily extend this to include other rules.

```python
import cv2
import numpy as np
import random
def apply_rule_based_augmentations(image):
    """
    Applies rule-based augmentations to the input image.
    Args:
        image: The input image as a NumPy array.
    Returns:
        The augmented image as a NumPy array.
    """
    # 1. Random Rotation (between -10 and 10 degrees)
    angle = random.uniform(-10, 10)
    rows, cols = image.shape[:2]
    M = cv2.getRotationMatrix2D((cols / 2, rows / 2), angle, 1)
    image = cv2.warpAffine(image, M, (cols, rows))
    # 2. Random Brightness Adjustment (between 0.9 and 1.1)
    brightness_factor = random.uniform(0.9, 1.1)
    image = cv2.convertScaleAbs(image, alpha=brightness_factor, beta=0)
    # 3. Random Horizontal Flip
    if random.random() < 0.5:
        image = cv2.flip(image, 1) # Flip horizontally
    return image
# Example Usage:
# Assuming 'image' is a NumPy array representing the input image
augmented_image = apply_rule_based_augmentations(image)
```

In this code, a random angle between -10 and 10 degrees is generated using `random.uniform()`. `cv2.getRotationMatrix2D()` calculates the rotation matrix. `cv2.warpAffine()` applies the rotation transformation to the image. A random brightness factor between 0.9 and 1.1 is generated. `cv2.convertScaleAbs()` adjusts the image brightness by multiplying each pixel value by the brightness factor. A random number between 0 and 1 is generated. If the random number is less than 0.5, the image is flipped horizontally using `cv2.flip()`. This framework provides a foundation for implementing various rule-based augmentation strategies and exploring their impact on model performance.

6.2.2 Attention-Based Augmentation

Attention mechanisms, initially popularized in natural language processing, have found applications in various domains, including image augmentation [14]. In this context, attention mechanisms can be used to:

Focus on Important Image Regions: Attention mechanisms can be effectively employed to identify and selectively augment specific regions of an image, ensuring that crucial features are preserved while allowing for more aggressive transformations in less critical areas [15]. In object detection, an attention module (e.g., a convolutional neural network) can be used to generate an attention map for the image. This map highlights the regions of the image that are most likely to contain objects of interest. Based on the attention map, the augmentation process can be selectively applied. In regions with high attention scores (i.e., regions likely to contain objects), more aggressive augmentations such as rotations, scaling, and cropping can be applied. This helps the model learn the robust features of the objects. In regions with low attention scores, milder or no augmentations can be applied. This helps to preserve the overall image context and prevent the distortion of important background information. This approach ensures that crucial object features are preserved while allowing for more aggressive data augmentation in regions that are less critical for object detection. This can meaningfully improve the model's ability to detect and localize objects accurately. For example, in an image containing multiple objects, the attention mechanism might focus on the regions around the objects. The augmentation process could then apply stronger rotations and zooms to these regions while applying milder augmentations or no augmentations to the background. This would expose the model to a wider range of variations in object appearance while preserving the overall context of the image. By selectively applying augmentations based on attention maps, models can learn more robust and discriminative features, leading to improved performance in object detection and other computer vision tasks.

Adapt Augmentation to Image Content: Attention mechanisms can dynamically adjust the intensity and type of augmentations based on the content of the image [16]. For instance, in medical imaging, regions with high image gradients often correspond to important anatomical structures or regions of interest (e.g., edges of organs, lesions). By applying stronger augmentations to these regions, the model is encouraged to learn robust features from the most informative parts of the image. In a chest X-ray, the model might apply more aggressive rotations or zooms around regions with high image gradients, which could correspond to lung nodules or other abnormalities. Conversely, in regions with low image gradients (e.g., homogeneous background areas), milder or no augmentations can be applied. This helps to preserve the overall image context and prevent the distortion of important anatomical information. The intensity and type of augmentations can be dynamically adjusted based on the specific characteristics of the image. For example, in images with high levels of noise, the model might apply less aggressive noise augmentation to avoid further degrading image quality. In a brain MRI scan, an attention mechanism could identify regions with high image gradients, which often correspond to brain structures or lesions. Stronger augmentations, such as small rotations or intensity variations, could then be applied to these regions to challenge the model and improve its ability to detect subtle abnormalities. By dynamically adjusting the augmentation strategy based on image content, attention mechanisms can significantly improve the effectiveness and efficiency of the data augmentation process, leading to more robust and accurate models in medical imaging and other domains. The entire process, including the attention mechanism, the augmentation policy, and the main model, can be trained end-to-end. This allows the model to jointly learn optimal attention weights and augmentation policies that maximize model performance.

Learn Augmentation Policies: Attention mechanisms can be integrated into deep learning models to learn optimal augmentation policies [3]. Instead of manually defining augmentation rules, the model itself learns to "attend" to specific features within the image and apply augmentations that are most likely to improve performance. The model learns to allocate attention weights to various regions or features of the image. These weights indicate the importance of each region for the specific task. For example, in object detection, the model might learn to allocate higher attention weights to regions containing objects of interest. The augmentation policy is then determined based on the attention weights. Stronger augmentations, such as rotations, zooms, or crops, can be applied to regions with higher attention weights, while milder augmentations or no augmentations can be applied to regions with lower attention weights. This ensures that crucial features are preserved while allowing for more aggressive transformations in areas that are less critical for the task. By incorporating attention mechanisms into the augmentation process, we can develop more intelligent and adaptive augmentation strategies that expressively enhance the performance and interpretability of deep learning models. This approach signifies a substantial improvement in data augmentation, moving beyond manually defined rules toward more sophisticated and data-driven methods for optimizing the augmentation process.

The benefits of Attention-Based Augmentation concerning interpretable image augmentation are as follows: (1) Improved Performance: By focusing augmentations on relevant regions or adapting to image content, attention mechanisms can improve model performance and generalization. (2) Increased Interpretability: Attention mechanisms can provide an understanding of which image regions are most important for the model's learning process. This can help researchers understand the model's decision-making process and identify potential biases. (3) Data Efficiency: By selectively applying augmentations, attention-based methods can potentially improve data efficiency by avoiding unnecessary or counterproductive augmentations.

6.2.3 Counterfactual Augmentation

Counterfactual augmentation provides a powerful tool for understanding and improving the interpretability of deep learning models [17]. Counterfactual Augmentation draws inspiration from the concept of counterfactual reasoning in causal inference, where we ask "what if" questions. In the context of image augmentation, it aims to generate synthetic images that explore scenarios that could have plausibly occurred in the real world but are underrepresented or absent in the original dataset. Many datasets exhibit biases, such as limited representation of certain groups or skewed distributions. For example, a facial image dataset might be biased toward lighter skin tones. Counterfactual Augmentation aims to mitigate these biases by generating synthetic images that counteract these existing biases. In the case of facial images, this could involve generating synthetic images with a wider range of skin tones, including darker skin tones that are underrepresented in the original dataset. If the dataset lacks images of certain objects or conditions, counterfactual augmentation can generate synthetic images that explore these underrepresented scenarios. For example, if a dataset of self-driving car images lacks images of rare weather conditions (e.g., heavy rain, snow), counterfactual augmentation could generate synthetic images that simulate these conditions by modifying existing images to incorporate realistic weather effects.

Overall the working principle of this technique is as follows [18, 19].

Identify Biases: Analyze the existing dataset to identify potential biases. This might involve analyzing the distribution of sensitive attributes (e.g., gender, ethnicity, age) and identifying any underrepresented groups or skewed distributions.

Define Counterfactual Scenarios: Based on the identified biases, define plausible "what-if" scenarios. For example, if a facial recognition dataset is biased toward lighter skin tones, a counterfactual scenario might involve generating synthetic images with darker skin tones while maintaining other facial features Another example is if a dataset of self-driving car images lacks images of rainy conditions, a counterfactual scenario might involve generating synthetic images with simulated rain effects.

Generate Synthetic Data: Employ generative models like GANs (Generative Adversarial Networks) or VAEs (Variational Autoencoders) to generate synthetic images that reflect the defined counterfactual scenarios. These models learn the underlying data distribution and can generate new images that are plausible and realistic.

Augment Training Data: Incorporate the generated counterfactual images into the training dataset. This expands the diversity of the training data and exposes the model to scenarios that were previously underrepresented.

By observing how the model reacts to counterfactual examples, we can understand its sensitivity to specific features or attributes. For example, if a facial recognition model consistently misclassifies images with darker skin tones, generating counterfactual images with varying skin tones can help identify and address this bias. Counterfactual examples can reveal limitations in the model's generalization ability. If the model struggles to accurately classify counterfactual images, it might indicate that the model is over-relying on specific features or is not robust to certain types of variations. By exposing the model to counterfactual examples during training, we can encourage it to learn more robust and generalizable representations. This can aid the model improved handling unseen data and improve its performance in real-world scenarios. Counterfactual examples can be used to debug model behavior. If the model consistently misclassifies specific types of counterfactual images, it can provide valuable clues about potential biases or errors in the model's decision-making process.

The following code snippet provides a basic framework for generating counterfactual images using a GAN. This is a simplified example, and we'll need to adapt it to our specific dataset, task, and requirements.

```python
import tensorflow as tf
from tensorflow.keras.layers import *
from tensorflow.keras.models import Model
import numpy as np
# Define a simple GAN for generating counterfactual images (simplified
example)
class GAN(tf.keras.Model):
    def __init__(self):
        super(GAN, self).__init__()
        self.generator = tf.keras.Sequential([
            Dense(128, input_shape=(100,)),
            LeakyReLU(alpha=0.2),
            Dense(256),
            LeakyReLU(alpha=0.2),
            Dense(512),
            LeakyReLU(alpha=0.2),
            Dense(784, activation='tanh')
        ])
        self.discriminator = tf.keras.Sequential([
            Dense(256, input_shape=(784,)),
            LeakyReLU(alpha=0.2),
            Dense(128),
            LeakyReLU(alpha=0.2),
            Dense(1, activation='sigmoid')
        ])
    def compile(self, discriminator_optimizer, generator_optimizer,
loss_fn):
        super(GAN, self).compile()
        self.discriminator_optimizer = discriminator_optimizer
        self.generator_optimizer = generator_optimizer
        self.loss_fn = loss_fn
    def train_step(self, real_images):
        noise = tf.random.normal((real_images.shape[0], 100))
        with tf.GradientTape() as tape:
            generated_images = self.generator(noise)
            real_output = self.discriminator(real_images)
            fake_output = self.discriminator(generated_images)
            # Discriminator loss
            real_loss = self.loss_fn(tf.ones_like(real_output), real_output)
```

```
                        fake_loss  =  self.loss_fn(tf.zeros_like(fake_output),
fake_output)

            discriminator_loss = 0.5 * (real_loss + fake_loss)


        gradients_of_discriminator = tape.gradient(discriminator_loss,
                                self.discriminator.trainable_variables)
        self.discriminator_optimizer.apply_gradients(zip(gradients_of_disc
riminator, self.discriminator.trainable_variables))

        with tf.GradientTape() as tape:
            generated_images = self.generator(noise)
            fake_output = self.discriminator(generated_images)
                generator_loss  =  self.loss_fn(tf.ones_like(fake_output),
fake_output)

        gradients_of_generator = tape.gradient(generator_loss,
                                    self.generator.trainable_variables)
        self.generator_optimizer.apply_gradients(zip(gradients_of_generato
r, self.generator.trainable_variables))

        return {"d_loss": discriminator_loss, "g_loss": generator_loss}


# Example Usage
gan = GAN()
gan.compile(
    discriminator_optimizer=tf.keras.optimizers.Adam(learning_rate=0.0002)
,
    generator_optimizer=tf.keras.optimizers.Adam(learning_rate=0.0002),
    loss_fn=tf.keras.losses.BinaryCrossentropy(from_logits=True)
)
# Train the GAN (This is a simplified example)
epochs = 100
for epoch in range(epochs):
    for image_batch in dataset:
        loss_dict = gan.train_step(image_batch)
    print(f"Epoch: {epoch}, D_loss: {loss_dict['d_loss']:.4f}, G_loss:
{loss_dict['g_loss']:.4f}")
# Generate counterfactual images
noise = tf.random.normal((10, 100))
counterfactual_images = gan.generator(noise)
```

6.2.4   Human-In-The-Loop Augmentation

Human-in-the-loop augmentation involves a collaborative approach where human expertise is integrated into the data augmentation process [20–22]. This synergistic approach uses the strengths of both humans and machines to create more effective and meaningful augmentations.

Human experts, with their deep understanding of the domain, can define specific rules and constraints for augmentation. For instance, in medical imaging, radiologists can define rules for simulating realistic patient movements, variations in imaging equipment, or changes in tissue properties. These rules ensure that the augmentations are not only diverse but also plausible and relevant to the specific domain. For example, radiologists can define acceptable ranges for small rotations (e.g., ± 5 degrees) that mimic minor patient movements during image acquisition. They can specify acceptable ranges for small translations (e.g., a few pixels) to simulate slight shifts in patient positioning. These rules ensure that the simulated movements are realistic and do not introduce significant distortions or artifacts that would not occur in real-world scenarios. Radiologists can define rules for adjusting image intensity, contrast, and sharpness within realistic ranges observed in clinical practice. They can specify the types and levels of noise that are commonly observed in images acquired from different imaging modalities and equipment. This ensures that the model is robust to noise variations encountered in real-world clinical settings. In cardiac imaging, radiologists can define rules for simulating variations in heart rate and cardiac cycle, which can affect image appearance. In lung imaging, rules can be defined to simulate respiratory motion artifacts, such as blurring or motion artifacts in the images. By incorporating this domain-specific knowledge, radiologists can guide the augmentation process to generate more realistic and clinically relevant training data. This ensures that the model is exposed to a wider range of variations that it is likely to encounter in real-world clinical practice, ultimately improving its diagnostic accuracy and robustness.

Humans can select or prioritize specific augmentation methods based on their understanding of the data and the task. For example, in a facial recognition task, experts might prioritize augmentations that simulate variations in lighting (Changes in ambient lighting, shadows, and glare significantly affect facial appearance), pose (Variations in head pose (yaw, pitch, roll), head tilt, and expression can dramatically change facial features), occlusions (Partial occlusions due to hair, glasses, or accessories are common in real-world scenarios) and facial expressions, as these are crucial factors in real-world scenarios. By prioritizing relevant augmentations, experts can avoid applying unnecessary or irrelevant transformations that might not improve model performance or even introduce noise. For example, applying extreme distortions or unrealistic color variations might not be beneficial and could even hinder the model's capability to learn expressive features.

Human feedback plays a critical role in refining the augmentation procedure. By actively reviewing augmented images, human experts can provide valuable insights and ensure that the generated data remains pertinent, meaningful, and of high quality.

Humans can assess the visual quality of augmented images, identifying and flagging any artifacts, distortions, or unrealistic features introduced by the augmentation process. For instance, in medical imaging, radiologists can identify augmentations that introduce unrealistic anatomical distortions or artifacts that might mislead the model. Human experts can evaluate the realism of augmented images. In the context of facial images, for example, humans can assess whether the augmented images still maintain a realistic appearance, including plausible skin tones, facial expressions, and lighting conditions. Human experts can assess whether the augmented images are appropriate for the specific task at hand. For instance, in a self-driving car application, humans can evaluate whether augmented images accurately simulate real-world driving scenarios, such as changes in weather conditions, traffic patterns, and pedestrian behavior. Humans can identify and flag any biases introduced by the augmentation process. For example, if an augmentation technique consistently produces images that favor a particular subgroup, human reviewers can identify and address these biases. The feedback provided by human experts can be used to iteratively refine the augmentation process. Based on human feedback, researchers can adjust augmentation parameters, modify rules, or explore new augmentation techniques to improve the relevance and quality of the augmented data. By incorporating human feedback into the augmentation loop, researchers can confirm that the generated data is of high quality, realistic, and appropriate for the specific task. This iterative process of human evaluation and enhancement leads to continuous upgrading in the augmentation process and ultimately enhances the performance and reliability of the trained models.

Human feedback forms a crucial part of an iterative refinement process. An initial augmentation pipeline is designed, often based on preliminary research and domain knowledge. This pipeline might include a set of basic augmentations (e.g., rotations, flips, brightness adjustments) with default parameters. A subset of augmented images is presented to human experts (e.g., domain specialists, and annotators) for evaluation. Experts provide feedback on various aspects. "Are the augmented images visually appealing and free from artifacts?". "Do the augmentations reflect realistic variations that could occur in the real world?". "Are the augmentations relevant to the task and do they improve model performance?". Experts might identify specific issues, such as unrealistic distortions, excessive noise, or biases introduced by certain augmentations. Based on human feedback, the augmentation pipeline is refined. This might involve: (1) Adjusting parameters: Modifying the range of values for parameters such as rotation angles, scaling factors, or noise levels, (2) Adding or removing augmentations: Including new augmentation techniques or removing those that are deemed ineffective or detrimental, (3) Modifying the order of augmentations: Changing the sequence in which augmentations are applied to achieve better results, (4) Addressing identified issues: Addressing specific issues raised by human evaluators, such as removing unrealistic artifacts or mitigating biases. The refined augmentation pipeline is then used to generate a new set of augmented images, which are again evaluated by human experts. This process of evaluation, refinement, and re-evaluation continues iteratively until the augmentation pipeline generates high-quality, realistic, and effective augmented data. This iterative refinement procedure permits researchers to continuously enhance the augmentation pipeline based on human feedback, confirming that the generated data is of the highest quality and best supports the learning process.

Human-in-the-loop augmentation often involves the use of interactive tools and platforms that facilitate human–machine collaboration. These tools can include:

Interactive annotation tools: Interactive annotation tools play a vital role in Human-in-the-Loop Augmentation by empowering human annotators to easily interact with images, define regions of interest, and apply augmentations directly [23]. Interactive annotation tools empower human annotators by providing a user-friendly interface for various image annotation tasks. These tools enable annotators to: (1) Easily draw rectangular boxes around objects of interest for tasks like object detection. This is a common and efficient method for defining the spatial extent of objects. (2) Create more complex shapes or outlines for objects with irregular boundaries, providing more accurate annotations for objects with intricate shapes. (3) Define pixel-level masks to precisely segment objects from the background, enabling accurate pixel-wise labeling for tasks like image segmentation. (4) Annotators can assign meaningful labels or categories to different objects or regions within an image, providing crucial ground truth information for training and evaluating machine learning models. Many tools like LabelImg, CVAT (Computer Vision Annotation Tool), Labelbox, etc. can directly interact with the image, applying transformations such as rotations, translations, scaling, shearing, and flipping using intuitive interfaces like sliders, drag-and-drop controls, or even touch gestures. This allows for immediate visual feedback and fine-grained control over the augmentation process. Many tools can adjust the parameters of augmentations (e.g., rotation angle, scaling factor, shear intensity) in real-time. This allows them to experiment with different augmentation strengths and observe the effects on the image directly, confirming that the chosen parameters produce realistic and meaningful transformations. Annotators can iteratively adjust augmentation parameters and observe the results, fine-tuning the transformation until the desired effect is achieved. This iterative process allows for more precise control over the augmentation process and ensures that the augmented images meet specific quality and realism criteria. Some tools can quickly identify and flag any artifacts or distortions introduced by the augmentation process. For example, excessive blurring, unnatural color shifts, or the introduction of unrealistic shadows can be easily spotted by visually comparing the original and augmented images. This allows for immediate identification and correction of issues with the augmentation pipeline, ensuring that the generated data remains of high quality and does not introduce noise or misleading information. By comparing original and augmented images, annotators can assess the realism of the augmentations. In medical imaging, for example, radiologists can easily identify if an augmented image exhibits unrealistic anatomical features or artifacts that would not be observed in real clinical scans. This ensures that the augmented data remains within the bounds of clinical plausibility and does not introduce unrealistic variations that could mislead the model. By observing the effects of different augmentations on the original image, tools can provide valuable feedback for iterative refinement of the augmentation pipeline.

They can suggest adjustments to parameters, identify problematic augmentations, and propose new augmentation techniques based on their visual observations.

Visualization tools: Visualization tools play a vital role in Human-in-the-Loop Augmentation by permitting human annotators to easily visualize and compare original images with their augmented counterparts [24]. This visual feedback is invaluable for several reasons. By visually comparing the original and augmented images, annotators can quickly identify potential issues such as unnatural warping, excessive blurring, or unrealistic color shifts that can be easily spotted. Augmentations might inadvertently obscure or remove crucial details within the image. Augmentations might introduce unintended artifacts, such as noise, jagged edges, or unnatural patterns. The visual comparison allows us to assess the realism of the augmented images. In medical imaging, for example, radiologists can easily identify if an augmented image exhibits unrealistic anatomical features or artifacts that would not be observed in real clinical scans. By visually comparing the original and augmented images, we can evaluate how effectively the augmentations are challenging the model and improving its robustness. For example, in object detection, they can assess whether the augmentations are successfully increasing the diversity of object appearances and poses. Based on their visual observations, annotators can provide valuable feedback for refining the augmentation process. They can suggest adjustments to parameters, identify problematic augmentations, and propose new techniques to improve the quality and realism of the augmented data.

Feedback mechanisms: Feedback mechanisms are crucial for effectively incorporating human expertise into the data augmentation process [25]. These mechanisms allow humans to offer meaningful understandings of the quality and efficiency of augmented data, guiding the refinement of the augmentation pipeline. Simple rating systems (e.g., 1–5 star ratings) can be used to assess the overall quality of augmented images. Annotators can rate images based on criteria such as realism, naturalness, and whether the augmentation preserves important features. These ratings can be used to prioritize and select the most successful augmentation strategies. Free-text feedback allows us to provide free-text feedback allows for more nuanced and detailed input. Annotators can describe specific issues with augmented images, such as unrealistic distortions, loss of important information, or the introduction of artifacts. They can also suggest improvements to the augmentation process, such as adjusting parameters, adding new augmentations, or refining existing ones. Platforms like Amazon Mechanical Turk can be used to efficiently collect feedback from a large number of human annotators. This can deliver a valuable understanding of the quality and efficacy of augmented data from a diverse range of perspectives. Some tools allow annotators to directly interact with augmented images, providing more granular feedback. For example, they might be able to mark specific regions of an image where the augmentation has introduced artifacts or distortions. By incorporating these feedback mechanisms, researchers can effectively use human expertise to refine the augmentation process, identify and address potential issues, and ultimately create higher-quality, more effective, and more reliable augmented datasets. This iterative feedback loop is essential for continuous improvement in the human-in-the-loop augmentation process.

By incorporating human expertise, Human-in-the-Loop Augmentation can confirm that the augmented data is of high quality, realistic, and relevant to the task. The involvement of humans can improve the interpretability of the augmentation process, as the rationale behind each augmentation decision becomes more transparent. By involving human experts, we can build greater trust in the augmented data and the models trained on it. Human-in-the-loop augmentation offers a promising approach for creating high-quality, interpretable, and trustworthy augmented data. By combining the strengths of human expertise and machine intelligence, we can develop more effective and reliable deep learning models that address the unique challenges of different applications.

6.3 Summary

This chapter explores cutting-edge advancements in image augmentation. AutoAugment, a pioneering approach, employs Reinforcement Learning to discover optimal augmentation policies directly from the data. It learns to select appropriate operations, their probabilities, and magnitudes, leading to highly effective and task-specific augmentations. Interpretable Augmentation focuses on enhancing transparency and understanding. Rule-based augmentation utilizes explicit rules, enabling control and domain knowledge incorporation. Attention-based augmentation uses attention mechanisms to focus on important regions, adapt to image content, and even learn optimal policies. Counterfactual Augmentation addresses biases by generating synthetic data that explore "what-if" scenarios, improving model robustness and fairness. These advanced techniques push the boundaries of image augmentation, enabling the development of more effective, efficient, and interpretable deep learning models. Human-in-the-loop augmentation uses the combined power of humans and machines to create high-quality and meaningful augmented data. Human experts play a crucial role in defining domain-specific rules and selecting appropriate augmentation techniques, ensuring that the augmentations are realistic and relevant to the task. Interactive annotation tools empower human annotators with the ability to directly manipulate images, apply augmentations, and observe their effects in real-time. Visualization tools facilitate the identification of potential issues and artifacts in augmented images. Furthermore, feedback mechanisms, such as rating systems and free-text feedback, enable humans to deliver valuable understandings of the quality and efficacy of the augmented data. This iterative procedure of human evaluation and refinement allows for continuous improvement of the augmentation pipeline, leading to more robust, reliable, and interpretable deep learning models. By combining human expertise with machine intelligence, Human-in-the-Loop Augmentation addresses the unique challenges of different applications and fosters the development of more trustworthy and effective artificial intelligence systems.

References

1. Sengupta, S., et al. (2020). A review of deep learning with special emphasis on architectures, applications and recent trends. *Knowledge-Based Systems, 194*, 105596.
2. Yang, Z., Sinnott, R. O., Bailey, J., & Ke, Q. (2023). A survey of automated data augmentation algorithms for deep learning-based image classification tasks. *Knowledge and Information Systems, 65*(7), 2805–2861.
3. Cubuk, E. D., Zoph, B., Mane, D., Vasudevan, V., & Le, Q. V. (2019). Autoaugment: Learning augmentation strategies from data. In *Proceedings of the IEEE/CVF Conference on Computer Vision and Pattern Recognition* (pp. 113–123).
4. Laskin, M., Lee, K., Stooke, A., Pinto, L., Abbeel, P., & Srinivas, A. (2020). Reinforcement learning with augmented data. *Advances in Neural Information Processing Systems, 33*, 19884–19895.
5. Khare, O., Mane, S., Kulkarni, H., & Barve, N. (2024). LeafNST: An improved data augmentation method for classification of plant disease using object-based neural style transfer. *Discover Artificial Intelligence, 4*(1), 50.
6. Chowdhury, S. R., Tornberg, L., Halvfordsson, R., Nordh, J., Gustafsson, A. S., Wall, J., Westerberg, M., Wirehed, A., Tilloy, L., Hu, Z., & Sjöberg, J. (2019, November). Automated augmentation with reinforcement learning and GANs for robust identification of traffic signs using front camera images. In *2019 53rd Asilomar Conference on Signals, Systems, and Computers* (pp. 79–83). IEEE.
7. Ma, L., Ding, Y., Wang, Z., Wang, C., Ma, J., & Lu, C. (2021). An interpretable data augmentation scheme for machine fault diagnosis based on a sparsity-constrained generative adversarial network. *Expert Systems with Applications, 182*, 115234.
8. Maharana, K., Mondal, S., & Nemade, B. (2022). A review: Data pre-processing and data augmentation techniques. *Global Transitions Proceedings, 3*(1), 91–99.
9. Kierner, S., Kucharski, J., & Kierner, Z. (2023). Taxonomy of hybrid architectures involving rule-based reasoning and machine learning in clinical decision systems: A scoping review. *Journal of Biomedical Informatics, 144*, 104428.
10. Cui, Y., Hanyu, E., Pedrycz, W., & Li, Z. (2019). Augmentation of rule-based models with a granular quantification of results. *Soft Computing, 23*, 12745–12759.
11. Shao, Q. (2024). Rule-based data augmentation for document-level medical concept extraction.
12. Berge, G. T., Granmo, O. C., Tveit, T. O., Ruthjersen, A. L., & Sharma, J. (2023). Combining unsupervised, supervised and rule-based learning: The case of detecting patient allergies in electronic health records. *BMC Medical Informatics and Decision Making, 23*(1), 188.
13. Hassija, V., et al. (2024). Interpreting black-box models: A review on explainable artificial intelligence. *Cognitive Computation, 16*(1), 45–74.
14. Chen, Y., et al. (2022). Generative adversarial networks in medical image augmentation: A review. *Computers in Biology and Medicine, 144*, 105382.
15. Jawahar, M., Anbarasi, L. J., Narayanan, S., & Gandomi, A. H. (2024). An attention-based deep learning for acute lymphoblastic leukemia classification. *Scientific Reports, 14*(1), 17447.
16. Obeso, A. M., Benois-Pineau, J., Vázquez, M. S. G., & Acosta, A. Á. R. (2022). Visual vs internal attention mechanisms in deep neural networks for image classification and object detection. *Pattern Recognition, 123*, 108411.
17. Verma, S., Boonsanong, V., Hoang, M., Hines, K., Dickerson, J., & Shah, C. (2024). Counterfactual explanations and algorithmic recourses for machine learning: A review. *ACM Computing Surveys, 56*(12), 1–42.
18. Temraz, M., & Keane, M. T. (2022). Solving the class imbalance problem using a counterfactual method for data augmentation. *Machine Learning with Applications, 9*, 100375.
19. Liu, Q., Kusner, M., & Blunsom, P. (2021, June). Counterfactual data augmentation for neural machine translation. In *Proceedings of the 2021 Conference of the North American Chapter of the Association for Computational Linguistics: Human Language Technologies* (pp. 187–197).
20. Kumar, S., Datta, S., Singh, V., Datta, D., Singh, S. K., & Sharma, R. (2024). Applications, Challenges, and Future Directions of Human-in-the-Loop Learning. *IEEE Access*.

21. Wu, X., Xiao, L., Sun, Y., Zhang, J., Ma, T., & He, L. (2022). A survey of human-in-the-loop for machine learning. *Future Generation Computer Systems, 135*, 364–381.

22. Yau, K. L. A., Saleem, Y., Chong, Y. W., Fan, X., Eyu, J. M., & Chieng, D. (2024). The Augmented Intelligence Perspective on Human-in-the-Loop Reinforcement Learning: Review, Concept Designs, and Future Directions. *IEEE Transactions on Human-Machine Systems.*

23. Monarch, R. M. (2021). *Human-in-the-Loop Machine Learning: Active learning and annotation for human-centered AI.* Simon and Schuster.

24. Kerdvibulvech, C., & Li, Q. (2024, June). Empowering Zero-Shot Object Detection: A Human-in-the-Loop Strategy for Unveiling Unseen Realms in Visual Data. In *International Conference on Human-Computer Interaction* (pp. 235–244). Cham: Springer Nature Switzerland.

25. Teodorescu, M. H., Morse, L., Awwad, Y., & Kane, G. C. (2021). Failures of Fairness in Automation Require a Deeper Understanding of Human-ML Augmentation. *MIS quarterly, 45*(3).